Technologies of the Human Corpse

Technologies of the Human Corpse

John Troyer

The MIT Press
Cambridge, Massachusetts
London, England

This book was set in ITC Stone Serif Std and ITC Stone Sans Std by Toppan Best-set Premedia Limited. Printed and bound in the United States of America.

Library of Congress Cataloging-in-Publication Data

Names: Troyer, John, 1972- author.
Title: Technologies of the human corpse / John Troyer.
Description: Cambridge, Massachusetts : The MIT Press, [2020] | Includes bibliographical references and index.
Identifiers: LCCN 2019028374 | ISBN 9780262043816 (hardcover)
Subjects: MESH: Mortuary Practice—history | Technology—history | Attitude to Death | Funeral Rites—history | Cadaver | Thanatology—history
Classification: LCC RA622 | NLM WA 11.1 | DDC 363.7/5—dc23
LC record available at https://lccn.loc.gov/2019028374

10 9 8 7 6 5 4 3 2 1

For my younger sister, Julie, and all the times she asked me if I was ever going to finish this book.

And for my parents, Ron and Jean, who also asked me the same question.

Lots of love to all three of you.

Contents

Acknowledgments

The following individuals, groups, and institutions have made the completion of this book possible: first and foremost Gracia Maria Redondo Nevado; my parents, Ron and Jean Troyer; my sister, Julie Troyer; the Department of Social and Policy Sciences and the Centre for Death and Society (CDAS) at the University of Bath; all of my CDAS colleagues, past and present; Matthew Browne and everyone at the MIT Press; Joanna Ebenstein and the Morbid Anatomy Library; Hayley Campbell for intervening when I really needed it; Caitlin Doughty; Phil Olson; Ara Francis; Lyn Lofland; All my Death World friends and colleagues; John Archer; Haidee Wasson; Cesare Casarino; Michal Kobialka; the Department of Cultural Studies and Comparative Literature at the University of Minnesota; Julietta Singh from the Department; all my students in all my classes I have ever taught; Kathie Walczak and the National Funeral Directors Association; Katherine Chew at the University of Minnesota Bio-Medical Library; Michael LuBrant in the University of Minnesota Mortuary Science Program; Taylor & Francis (http//:taylorandfrancis.com) for originally publishing chapter 1 as John Troyer, "Embalmed Vision," *Mortality* 12 no. 1 (2007):

22–47, and chapter 3 as John Troyer, "Technologies of the HIV/ AIDS Corpse," *Medical Anthropology* 29, no. 2 (2010): 129–149; Mary Roach; Candi Cann; Kristi Bakken; Ari Hoptman and Val Pakis for help with my Greek language questions; Amber Tribe; all of my grandparents who died many years before this book came out but who remain at my back at all times; Carol Troyer; Keith Troyer; my amazing and indispensable high school teachers at Hudson High School in Wisconsin; the Chupacabra; and the Bisga Man.

Preface

I needed to finish this book before my entire family died.

While writing, rewriting, and then editing what you are now reading, my younger sister, Julie Troyer, was diagnosed in late July 2017 with an aggressive glioblastoma multiforme brain cancer and underwent emergency surgery to reduce the tumor's size. She then received multiple radiation and chemo treatments for the next year before dying on July 29, 2018, in Milan, Italy, where she lived with her family and worked as a schoolteacher. She is survived by her husband and two young children.

In addition to my sister's death, my father's younger brother Keith died from multiple terminal conditions on June 15, 2018; the same night I received an Alumni Achievement Award from the University of Minnesota for my academic work in the University of Bath's Centre for Death and Society on end-of-life issues. Uncle Keith died, by chance, in a University of Minnesota Hospital room two blocks from the awards ceremony. I attended the ceremony with my parents, and we only learned that Keith had died after we left campus and drove thirty-five minutes home to Wisconsin. Following a quick turnaround, we drove back to the hospital and said goodbye.

My father, Ron Troyer, survived a series of heart attacks that started in late 2015 and subsequently required inserting twelve stents into his coronary arteries. As I type these words in early 2019, my parents are preparing to meet with his cardiologist to discuss scheduling open heart surgery, since everyone agrees the less-invasive route is no longer viable. My mother, Jean Troyer, is in relatively good health (thankfully) but does manage some severe arthritis and other age-related conditions.

As for me—I should really lose some weight given all the cancer and heart disease raging through my genes (I'm working on it). But given all these recent events, the past few years have fundamentally changed how I understand what living and dying mean for my family.

At this point I need to explain that not only am I director of the Centre for Death and Society at the University of Bath, one of the world's only research centers dedicated to the interdisciplinary study of death, dying, and the dead body, but that my father is also a retired American funeral director. When I say my entire life revolves around death, it is not hyperbole. Death is all I have ever known, and this book is a hybrid response to the death, dying, and dead body constellation that inhabits my life. It started as a mostly academic(ish) text but morphed with time into a book that asks readers to think about death, dying, and dead bodies in radically different ways. This book is also now unintentionally part memoir. I do not consider myself a memoirist, and I do not ever plan to write a memoir, but my sister's death interjected itself into these pages in ways I cannot ignore. I firmly believe that fighting death is a bad idea. Death wins. Death always wins.

Saying Dying

My supposed familiarity with death and dying is why I began this book with my sister; despite my lived experience and academic credentials on human mortality, I was completely unprepared for her untimely death at the age of forty-three.

But I was not unprepared in that way many people are wholly unprepared for a person to die. There was an element of that emotion, but I was raised to understand that any person, especially the people we know and love, unexpectedly die all the time. Death was totally normalized for me in this way. My self-described unpreparedness arrived when at almost every stage of Julie's one year living with brain cancer I felt compelled to intervene and ask about hospice care, to make sure that my sister understood she was dying. But for reasons I do not fully understand I did not say anything until the very end.

That is, I finally said something on the night of July 13, 2018, when I was with my sister in Italy and she asked me what was happening, so I told her she was dying. I held her hand and did what none of her medical team had done and said what the counselors kept telling my brother-in-law not to say—I told my kid sister that she was never going home to Wisconsin, that she was going to die in Italy, and that we would do everything we could to make her end of life comfortable. Julie died sixteen days later.

What I remember most about this entire experience is my sister saying three things: (1) thank you for telling me, (2) I know I'm dying (we did grow up the same way, after all), and (3) I am glad you are the one who told me because I would do the same for you.

This was one of the last conversations I ever had with my sister.

Julie's overall health was rapidly declining by this time. She had already lost vision in her right eye because of the tumor, the left eye was on its way out, she was bedbound because of severe balance problems also caused by the tumor, and her hearing was failing. On the upside, she was not in much physical pain because a local hospice recently started providing her with palliative drugs and outpatient medical care, complete with on-call doctors and nurses. But even after the local hospice got involved, no one said anything about dying. It was a matter of days after our conversation that an ambulance transported my sister to the same hospice's comprehensive inpatient facility for comfort care. Julie requested that she not die at home, contrary to the conventional wisdom on where people prefer to die, so my brother-in-law dutifully made the arrangements.

Two nights after I told my sister that she was dying, I asked if she wanted to talk on Skype with Mom and Dad in Wisconsin. She said yes, so I stretched out next to her on the bed and held up my laptop so we could all see each other. The conversation was short, mostly because it was difficult for my sister to maintain even medium-length, focused conversations by this point, but we did talk about how much we loved each other and how Julie understood she was dying and how unfair it all felt.

A common narrative around terminal illness often describes dying people walking peacefully toward death, exhibiting stoic acceptance. My sister was angry about dying. She fully understood and accepted that death was coming, there was no supposed death denial in her final weeks, but her exact quote to my parents was "Dying sucks. It really sucks." Julie just wanted a longer life. We did too.

Neither my parents nor I could know it at the time, but this was the last conversation all four of us ever had together, even if it was on my computer screen.

Writing a Death Book

My sister and I talked a lot about this book—about how long it took to get published after a series of disasters with one publisher and how MIT Press and my editor, Matthew Browne, came to my rescue. About how she called me the Overlord of Death. For years my father joked that I needed to finish the book soon, ideally before he died, so that he could actually read it. We all laughed until he had his first heart attack, at which point I realized finishing the book was no joke. Then my sister got sick and died before anyone expected.

I knew in April 2018 that Julie would never see this book. I was in Italy with my partner and parents for Julie's birthday, and it was abundantly clear even then that the cancer and its associated treatments were aggressively dismantling my sister. It was also during this trip that I started a series of prose pieces called *Watching My Sister Die*, some of which are included in this book. Looking back at April 2018, I made a mistake by not saying anything about end-of-life care and dying. But it was also Julie's birthday party, and my partner always reminds me that most of her friends were not ready to hear the word *dying*. My partner is absolutely correct, and I understand all this, but I still struggle with balancing what I professionally recognized versus what I personally said.

The whole situation felt bleakly ironic, since being a funeral director's kid meant death was our familiar friend. My sister and I spent hours and hours of our youth in the different funeral homes Dad worked in. We never actually lived in a funeral home, much to the disappointment of *Six Feet Under* fans worldwide, but we did grow up watching our father moving RIP floral arrangements around, or vacuuming a visitation chapel's carpet, and frequently disappearing on Christmas mornings because someone had died the night before. Normal. This was all totally normal for us. Julie and I both understood that our childhood was different than most, but it was all we ever knew. Indeed, our parents had made a point of not hiding anything about death, funerals, or dead bodies when we were kids. One of my earliest childhood funeral home memories (these are a thing—ask anyone who grew up around funeral homes) is touching the dead hand of an older woman before her visitation started. I remember her skin's distinct coolness as she lay in the casket, and asking my parents why the hand felt cold. They both explained that the woman was dead, and that a person's body temperature changed after dying. Her hand was completely normal, they said. Dead bodies were completely normal, they said. And this was when I learned, at a very early age, that human corpses were not scary— they were just dead and cool to the touch. I also remember the blue polyester pantsuit that the woman wore, mostly because this all happened in the mid-1970s.

When my sister took her final turn and died, I was in Bristol, England (where I live), preparing to board a 4:55 p.m. flight for Italy. This was on a Sunday. A mutual college friend, already at the hospice, called me on my phone while I sat waiting to board the plane and told me that Julie had died holding her husband's hand. The original plan had involved my flying to Italy later that week, but after some calls from my sister's husband and her friends over the weekend, we all agreed that I should get to Milan as soon as possible. My niece and nephew had said goodbye the day before. The same mutual friend texted me several times while I waited in the airport to say that Julie's breathing had become more labored, so I was not surprised when the final call came in. I made sure during all the predeath texting that our friend knew Julie did not want any resuscitation, should the hospice decide to take measures keeping her alive. Everyone around Julie agreed on this point, and it would be strange to see a hospice make such a move, but my brotherly protectiveness took hold and I needed to make sure that my sister remained dead if and when she died. The last thing I wanted was for the hospice to resuscitate my sister because they knew I was on the way. Communicating all this over text messages was surreal, but I also know it is how we humans die now. During that final phone call I kept repeating, Do not move her body. Keep her body there. I want to see Julie when I get to the hospice, and our friend said not to worry.

I called my parents. I called my partner, who was in the middle of Norway working as a tour guide with a group, so she could not join me in Italy. I told everyone that our friend would make sure the hospice kept Julie's body on-site so that I could see her when I finally arrived. I asked a young married couple with two children, who overheard my conversations and stared at me with extremely concerned and slightly shocked eyes, if they would stall for time before getting on the plane so that I could get my things together. It's okay, I assured the young mother. It's okay. I'm okay. I'm the Overlord of Death. Then I walked onto the tarmac and boarded the plane. Gate 11. Bristol Airport. If anyone wants to know exactly where I learned my sister died, I still have the boarding pass.

On that Sunday afternoon I could only book a one-way flight that took me from Bristol to Amsterdam and then Milan, so by the time I finally arrived at the hospice, my sister had been dead for several hours. The nursing staff had moved Julie from her room down to the hospice's *Mortuaria* level. I know this because I saw *Mortuaria* on the elevator signs as I descended to the basement, making a mental note along the way about experiencing the world's weirdest version of Dante's *Inferno*. Two extremely compassionate nurses showed me to the private room and told me in halting English to take as much time as I needed. I used my restaurant-level Italian to thank them.

Stainless steel. That is the first thing I noticed when I walked into the room. My dead sister's body laid out on a stainless steel table, next to a stainless steel counter complete with a sink and retractable hose. She still wore her hospice gown, and a shroud covered her lower torso. In the room's floor, at the foot of the table, sat a drain. It took me two seconds, using my advanced-degree critical thinking skills, coupled with growing up in funeral homes, to realize I was looking at my dead sister in the hospice's prep room.

The prep room is where a funeral director "prepares" a body for a funeral.

What I am about to say may not make sense, but I took immense comfort seeing my sister this way. In that room. On that stainless steel. In a mortuary.

This was our youth. We grew up in these rooms.

I was far more at ease talking with Julie in that familiar stainless steel room, holding her hand, hugging her, and kissing her goodbye, than if we had been in her hospice room. I spent around forty-five minutes with my sister, reading her what I had written on the different airplanes and making sure she knew that I would look after our parents.

It took me a long time to actually leave. I wanted nothing more than to sit for days and hold my dead sister's hand and think about all the prep rooms we saw as children.

Late that night, as I left the hospice with everyone, I noticed that the lampposts outside the building all had advertisements for local funeral homes, or *funebri* in Italian. Knowing "funeral home" or its equivalent in multiple languages is one of my many skills. And not small adverts, I mean large banners that were impossible to miss no matter how hard you tried. They reminded me of banners announcing Fourth of July parades or gay pride festivals, but designed by Goth kids who went into marketing. I asked my brother-in-law if this was normal in Italy, because, well, I did not think you would see banners advertising funeral homes outside US or UK hospices. He said it was normal(ish) and just proved that Italian businesses knew where to find their customers. Then we started laughing because we both knew that Julie would think the funeral home advertisements were funny and that I must tell Mom and Dad about them. Rest assured, I did. They eventually took photos of the banners.

I slept that night at my sister's house in the same bed where I had told her she was dying. Two days later, on July 31, I went with my brother-in-law to central Milan to file all the necessary death notification paperwork. It was a tough day, for all the obvious reasons, but especially difficult for him.

July 31 is his birthday.

August 1 was the funeral.

Ending with a Funeral

Julie's funeral service took place in an exceptionally stylish and modern Italian funeral home. A kind of funeral home very different than the midwestern American ones we knew as kids, but a funeral home all the same. She always appreciated marble floors and modernist architecture; the Jesus stuff (as we both called it) not so much, but, you know, this was Italy. Julie and I routinely joked about spotting funeral homes of any kind in any city before anyone else. "That's a funeral home," I would say or she would say, well before any visible signage appeared. Our friends would then look at us with slightly terrified bemusement, as if we saw Death silently creeping up behind them in a *New Yorker* cartoon. We just intuitively recognized these places.

My brother-in-law organized the funeral within hours of my sister's death, contacting the funeral home and meeting with someone while I made my way to Italy. Funeral planning is emotionally difficult no matter the circumstances, but I could tell that my brother-in-law needed/wanted to make these arrangements on his own out of love and devotion. Besides, had I gone with him to meet the funeral director, I would probably have spent more time asking questions about how and why Italian funeral homes did certain things than actually focused on my sister. He also made sure that the funeral home staff knew his deceased wife's father was a retired American funeral director, and I got the sense that the unspoken code many American funeral directors follow when helping a colleague's family also happened in Italy.

My parents flew in that day from Wisconsin, and we collected them at the airport on the way to the funeral. By chance, they had already planned to arrive in Milan on August 1 and had booked their tickets months in advance. When we gathered in late April for Julie's birthday, we hoped that she might live long enough for one last Troyer family visit come August; instead the four of us said our final goodbyes at the funeral home.

That last goodbye is when my familiarity with death, dying, dead bodies, funerals, funeral homes, funeral directors, grief and bereavement, the broad academic literature on when a loved one dies, and watching parents weep over their dead child became profoundly conflicted. I knew exactly what to do with my parents when we arrived at the funeral home and my death professional mode immediately kicked in. At the same time I remember thinking that this experience of death was never supposed to happen—that my younger sister was never supposed to die first and that our parents were never supposed to see Julie dead. Julie and I occasionally talked about who would die first, Mom or Dad. But we never *seriously* discussed what would happen if either one of us died first. I say seriously because I bought a motorcycle in my early thirties during a pre-midlife-crisis moment, which meant my sister told me in no uncertain terms that she would kill me if I died riding it. I thought about that motorcycle a lot after Julie died and how unfair it seemed that despite my best efforts I managed to live, while she died from brain cancer. Connecting the two is not wholly rational, but it is often the way a person thinks while holding a dead sibling's hand.

Mom and Dad spent a long time with Julie in the private visitation room where mourners can see the deceased before the actual funeral. Then it was time to move Julie's body, put her in the European-style tapered-end coffin selected by her husband, move the coffin to the funeral chapel for the service, and close the coffin's lid.

I spent time with my sister in the visitation room the day before (after filing the death notification paperwork), and she looked good, which was a relief because I did not want to call my parents and tell them that the funeral home handling their dead daughter botched the job. It seems crass, I know, to describe "botched jobs," but this was my immediate concern. If you grow up around funeral homes, then you know what poorly done body prep looks like. Julie and I certainly knew.

Curious death world readers will surely wonder whether my sister was embalmed; she was not. Embalming is not widely done by this specific funeral home (or across Italy, really), so she was placed on an electric cold pad that slowed her body's decomposition. The funeral home dressed her in the clothes provided by her husband, applied cosmetics, and styled her hair. Again, she looked good. In addition to funeral directing, my father also taught mortuary science for many years—specifically, funeral cosmetology—so trust me when I say that Mom, Dad, and I spent a long time scrutinizing how the funeral home prepared my sister for public viewing—in between crying and occasionally chuckling at the enormous Jesus crucifix above Julie's head, which she would never want.

Then the funeral happened. The neo-rococo-style chapel quickly filled to capacity, so people spilled out into the funeral home's foyer and quietly listened. The floral arrangements contained many sunflowers and gerbera daisies, my sister's favorites. Musician friends played music. And everyone told funny stories about Julie in both English and Italian, which many of the Italians found quite different from their usual, more somber funeral practices but very much liked. Julie was cremated the next day, and her urn now sits on a family bookshelf in Italy so that her children can say hello to Mama.

I argued with myself for a long time over how to describe my sister's funeral—in that way writers sit at a keyboard and talk to themselves, not realizing other people can hear them, saying, "Make sure and describe your dead sister's body but don't make it weird." A key reason for writing this book, and my entire career writ large, is to understand all the invisible technologies humans use to make modern death and dying visible. Defined. Knowable. Experienced. I could go on for pages, and let's be real—I do go on for pages. So, for example, when I saw the electric cold pad keeping my sister's body from decomposing, I automatically recalled the entire history of noninvasive preservation technologies used in funerals since the nineteenth century. But when it came to Julie's funeral, all I really wanted to say is that it happened. She lived. She died. Everyone laughed and cried. People instinctively posted things on social media. Language seemed impossible. Words were not applicable. And for years I read academic articles and books that quoted grieving people saying what I was now feeling, which meant I unintentionally prepared myself for Julie's death but also meant I simply wanted to say I get it now. I think I get it now.

Disaster

Eventually, this preface needed a terrible twist (more than just my sister dying), and disaster finally struck when I left the journal documenting everything about Julie's death and the last ten years of my life on the airplane back to England. I put the journal in the seat pocket after writing about Julie's funeral and then promptly forgot to collect it when the plane landed at Gatwick Airport. Twenty minutes into the train ride back to Bristol I reached into my backpack to retrieve the journal and jarringly realized my mistake, so I returned to Gatwick and begged the airline for access to the plane. That was never going to happen, but I did get a phone number to call and a website at which to register my lost item. The office that returns lost items was already closed for the day, so I left one of those voicemail messages that go on for so long that the system eventually cuts you off. I then called back and left a second, equally long message about my sister and her brain cancer and her death and how this journal meant more than anything to me and that I was a death studies academic (no idea why that was important to say, but I did) and that if you opened it up, you would find a series of poems called *Watching My Sister Die*, so PLEASE PLEASE PLEASE get the journal back to me. Thank you. Sorry this message is so long.

That was all I could do at 9:00 on a Monday night, so I got back on the train and headed home, filled with such self-directed rage that I felt numb. This was when Julie's death really hit me, when I sat by myself on the train unable to write down what I was feeling because my journal was in lockup at Gatwick Airport. This is also when I started talking to Julie, asking her to help me get the journal back. And if anyone ever asks—it is 100% normal to talk to dead people. I am not joking. Try it. The dead are exceptionally good listeners.

The next day after some more frantic phone calls and the serious shedding of tears, a lovely Scottish woman (whose name I cannot remember) calmed me down over the phone and personally made sure that my journal made it back to me. I rarely let the journal out of my sight now, and I am looking at it as I type these words. But for writing, I do not know how I would ever process my sister's death. I would find ways, I am sure, but like Bartleby, I would prefer not to.

At this point in the story, all the days and emotions begin to blur together. We had another standing-room-only memorial service a month later in Wisconsin, for which my father manually scanned over 150 photos documenting Julie's life and saved said photos in six different "Memories of Julie" desktop folders that I then turned into individual PowerPoint presentations and clicked through during the service. I must admit that using PowerPoint during my sister's memorial service caused me mixed emotions, but it did the job, and contemporary presentation software definitely has its historical place in twenty-first-century memorialization technology. It also felt good to be doing something useful during the service. My brother-in-law flew in from Italy, and my partner was able to join us this time. A friend from high school who is now a professional opera singer sang "Unforgettable"—a song he also sang with my sister for a high school choral concert many years ago. Other friends delivered eulogies. Another friend volunteered to do the floral arrangements. Two friends who could not make it that day recorded music, and we played the recording over a photomontage. We held the memorial service in a local hotel ballroom, and Wisconsin being Wisconsin, this meant an open bar after everything finished. My sister always liked an open bar.

I can wholeheartedly confirm that if the Troyer family does anything well, it is planning and running a funeral. This may or may not come as a surprise. My Mom oversaw the front of house greeters and overall hospitality, including, but not limited to, preparing cupcakes for several hundred people; I ran the audiovisual, and my Dad stage-managed the entire affair. But working together on the memorial service also made me sad because I do not know what I will do without my sister when my parents die. I just really want my sister at those funerals.

Dammit, Julie.

I always thought a preface to this book would chronicle the ways growing up around funeral homes made me see the world differently. Indeed, the initial preface opened with a mildly amusing anecdote about my father explaining how dead bodies are not actually cold—they are just room temperature. This really happened. I'm not making it up. But I only recently realized that how I grew up did something profoundly more important: it prepared me for the moment I became an older brother holding his terminally ill sister's hand and saying, Julie—you're dying.

And now I am an only child looking after his aging parents.

Like I said, I needed to finish this book before my entire family died.

Postscript

On November 29, 2019, my father, Ron Troyer, collapsed in the Minneapolis-St. Paul International Airport from what we now know was cardiac arrest. He and my mom had just returned that afternoon from a cruise in the Bahamas and were headed to the car rental counter. True to dramatic form, my dad collapsed at the top of a long escalator and fell backwards onto my mom, who managed to keep him from sliding all the way down the steps. My father is a big and tall man, so stopping his descent was no minor feat.

Members of the general public started CPR before an entire brigade of EMTs, fire fighters, and police officers arrived on the scene and restarted my dad's heart. The paramedics then took my father by ambulance to the St. Paul, Minnesota, intensive care unit that began treating him for heart problems in 2015.

Many thanks to the young woman who held my mother's hand during the whole resuscitation ordeal. The empathy of strangers is most often apparent in these kinds of crises.

As I type this on December 3, 2019, my father is still in the ICU, resting comfortably, and his vital signs are good. But he has never regained consciousness, and I do not think he will. His brain went without enough oxygen for too long after he collapsed. I will soon fly to the United States to help my mom follow my father's end-of-life care wishes and medical directives. I am grateful that we discussed all these possibilities a few years ago and my parents made their wishes known in writing. My father's Living Will is already out and sitting underneath my black journal.

Two things:

1. My dad knew this book was finished. He saw the proofs and cover art, and he told me how much it meant to him that I wrote about Julie in the preface. My parents visited England earlier in November, and one of my last conversations with my father was about how much seeing Julie in these pages meant to him. He and I didn't so much talk about the preface as just cry and nod; in that way, a father and son sometimes don't need words to communicate thoughts.

2. This is how my father wanted to die. Quick. Fast. Never regaining consciousness. After a fabulous cruise with my mom. In an airport. My dad really loved airports and airport lounges and everything about international travel. I'm the same way.

So, my dad isn't dead. Not yet. But I have a strong hunch it's just my mom and me now. And seeing my father suddenly fall into a comatose state so soon after watching my sister die is impossible to fully describe: I understand what is happening, yet I do not want to understand what is happening. I keep coming back to Samuel Beckett, "You must go on, I can't go on, I'll go on."

But right now I need to look after my mom. It's okay. I'm okay. I'm the Overlord of Death.

4/29/2018

Watching My Sister Die
I am too young to write these words
too immature
too consumed with my own self-doubt
and too knowledgeable about death   to see how and why
my sister will soon die.
Up against a ticking clock   that won't stop
and that will most certainly break my sister down   before
my parents die themselves.
So I'm   adrift   right now because I just spent
three days watching my sister enter the finality
of her life →
As we sang Happy Birthday and I struggled to even
finish the words.
knowing that this is/was   most likely her last party.

And still I can't find a way to write this all down.
To capture how much I cried and how many tears
I saw as everyone comes to understand what's really
happening.
Sitting in this second-rate airport lounge knowing that
my sister will never leave this city again.
That this is where she will die.
That in this same fucking ridiculous second-rate airport lounge
I held Santa Maria in my arms   when she could
no longer hold back the tears.
Trying so hard to be strong for the two of us.
And so I'm lost right now old friend.
    Staring out the window
Wanting to tell Death all about what's happening
knowing that eventually I will. But not now.
I burned that possibility in my haste and loss of control.
                        Realizing my   mistake
staring into the void.

A void I can see slowly consuming my quickly dying sister.
So I'm coming back to this page and words again
Old Friend
to the untimely death of my too-young-to-die-sibling
knowing that no matter how much I write and   how
much ink I spill
this death is going to happen far sooner than anyone understood.
And I will need you more than ever Old Friend
to let me run my thoughts sideways across your pages
so that I   continue to choose life
even as my sister deteriorates toward death.

Introduction: The Human Corpse

On July 9, 1981, the President's Commission for the Study of Ethical Problems in Medicine and Biomedical Behavioral Research released its report *Defining Death: A Report on the Medical, Legal, and Ethical Issues in the Determination of Death.*[1] The commission's work focused on questions surrounding the legal standing of human individuals kept artificially alive by machines. As the report's introduction explained, the mandate of the president's commission was to "study and recommend ways in which the traditional legal standards can be updated in order to provide clear and principled guidance for determining whether such bodies are alive or dead."[2] In America during the late 1970s and early 1980s new kinds of medical technologies were making death less absolute and producing an entirely new kind of "almost" or "nearly" dead human body. The president's commission confronted a significant dilemma: "Such artificially-maintained bodies present a new category for the law (and for society), to which the application of traditional means for determining death is neither clear nor fully satisfactory."[3]

What the commission's final report addressed, now over thirty-five years ago, was a complicated subject that remains pertinent today, namely, the ongoing redefinition of when a human body is considered dead for science, medicine, and the law. More broadly speaking, the commission asked, When does a body die? And when does it then become a dead body? This book responds to those questions by critically examining the dead human body's historical and theoretical relationship with technology. These historical and theoretical technologies encompass the physical machines, political concepts, human laws, and sovereign institutions that humans use to control the corpse by classifying, organizing, repurposing, and physically transforming the dead body every day.

Calling this book *Technologies of the Human Corpse* necessarily requires investigating the history of machines used to alter the dead body. It also means presenting and critiquing the intertwining of those same mechanical devices with human politics and understandings of death.[4] To write about the history, theory, and future of the corpse in the early twenty-first century necessitates stripping away the countless invisible machines visually and biologically controlling the dead body for human viewing. Simply pulling back the curtain on the mechanical technologies producing the modern human corpse, however, is not enough. The true challenge confronting this project is articulating how the technologies deployed to produce the human corpse are as much conceptual as material. These material technologies are discussed in a variety of ways and include nineteenth-century embalming machines, 1970s death groups, biohazard clothing for handling HIV/AIDS corpses, black-market sales of human tissues, and *Body Worlds* displays. The conceptual technologies are no less material and produce equally tangible results, for example, the institutional role of organizations such as the American funeral industry in standardizing the concept of a modern human corpse, the political obfuscation of detainee deaths in military internment camps run by the US government, and even the conceptual possibility of using patent law to contest and redefine the very idea of human death.

Building an argument about the conceptual and material technologies interfacing with the human corpse involves using the work of various critical theorists, cultural historians, and social scientists. But it also requires using the writings and textbooks of entirely forgotten nineteenth-century funeral directors whose work in human embalming established a technological paradigm that remains in use today. In its own modest way, this project presents an argument about how technologies developed by humans during the nineteenth century to control the appearance of human corpses have significantly contributed to the contemporary American (and increasingly First World) understanding of death and the dead body. Those earlier, nineteenth-century technologies often changed forms or simply disappeared. Yet their effectiveness at challenging the concept of human death persists today.

This entire project is deeply indebted to the historical and theoretical work of countless individuals who previously explored the limits of human bodies, long before I ever contemplated the politics of corpses. One of those individuals, philosopher Giorgio Agamben, made a simple but important statement that affected my own way of thinking about the dead body. In his book *Homo Sacer*, Agamben explains the difficulties in defining contemporary living and dying: "This means that today ... life and death are not properly scientific concepts but rather political concepts."[5]

Agamben's insight remains ever relevant, but these pages present more than just rehashed debates about defining death. This monograph on the human corpse contributes to the broader public debate about the politics of life and death in the modern Western world by beginning, not concluding, with the dead body. The broader goal is to make readers of diverse backgrounds think about their own human mortality by contemplating how death and the dead body produce numerous kinds of conflicting meanings, contested situations, and political struggles for the twenty-first century.[6] Just as the work of the president's commission contributed to that debate by parsing the meanings of life and death in an age of biomedical machines, I too want readers to think about the increasing, everyday overlap between technology, politics, dying, and the dead body. A fundamentally important first step toward that goal requires asking two key questions: What do you want done with your body when you die, and have you conveyed those wishes to your next of kin? It is also equally important for readers to contemplate these questions alongside the "future of death" and the dead body's long-term relationship with constantly evolving definitions of human mortality. How to legally understand the future of death and the dead body was a key question for the president's commission when it asked "... whether the law ought to recognize new means for establishing that the death of a human being has occurred?"[7] And that specific question remains relevant three decades later.

Both the commission's report and relevant scholarship on these topics demonstrate how death is exceedingly difficult to define given the quasi-living conditions in which human bodies can exist for years. Professor of Law and Bioethics Alta Charo describes the situation this way: "... choosing a legal definition of death entails deciding whether legal death should coincide with biological death and ... the biological definition of death is, like many biological phenomena, inherently ambiguous."[8] Added to these already ambiguous distinctions is the question of when a *person* is medically dead even though the person's body may still be legally and biologically alive. As Charo explains, "The word *person* is used as a term of art in law to signify an entity granted equal protection of the law. The term is *not* co-extensive with the biological concept of 'a live human'; corporations can be persons while fetuses or antebellum slaves are not."[9]

An even more complicated question to answer is this: When does a living person (broadly defined) become a corpse? In other words, when precisely is a person defined as totally dead? The 1980 Uniform Determination of Death Act (UDDA) largely answers this question, and it serves in large part as US common law for determining when an individual is in fact completely dead. The UDDA stipulates, "An individual who has sustained either (1) irreversible cessation of circulatory and respiratory functions, or (2) irreversible cessation of all functions of the entire brain, including the brain stem, is dead."[10] Pulling all of these strands together then suggests that a corpse is a human body with an irreversible absence of brain activity and/or cardiopulmonary reflexes which makes the legally distinct person a biologically dead body. And yet, defining the corpse and the postmortem conditions that surround the human body are never this simple or straightforward.

All of these definitions, conundrums, and ethical limits regarding death and the human corpse have not dissipated since the publication of the presidential commission's report in 1981. On the contrary, the arguments about what makes a dead body a no longer living person have only grown more complicated and rigorously managed by sovereign state authorities. Death and the dead body have never been more alive in public debate and imagination. One of the main causes of that public debate is the human body's relationship with modern medical technology. Numerous kinds of biomedical machines have produced twenty-first-century human bodies that are not allowed to die. In other words, life in the human body will be prolonged until the practice of living simply cannot be either biologically or financially sustained. *Everything must be done and at all costs to preserve life*, so one argument about using these technologies goes. The legal, biological, and medical definitions of contemporary death subsequently run into problems since "life" has been pushed beyond historically constituted human boundaries.[11] Agamben describes this situation another way, using the pioneering work of two French neurologists working on irreversible human comas in the early 1960s: "Mollaret and Goulon immediately realized the significance of *coma dépassé* far exceeded the technico-scientific problem of resuscitation: at stake was nothing less than a redefinition of death."[12]

The human corpse is the central figure that marks the boundaries of these debates about the limits of life and death: a quasi-subject, an almost-person, a body with some legal rights but not the same rights as a living person. As T. Scott Gilligan and Thomas F. H. Stueve explain in their book *Mortuary Law*, the human body effectively becomes quasi-property after dying:

> It is not property in the commercial sense, but the law does provide a bundle of rights to the next of kin in relation to that body. The survivor is given the right to take the body for purposes of disposition, to allow body parts to be used within the confines of the law, to exclude others from possession of the body, and to dispose of the body. This bundle of rights renders the dead body the quasi-property of the surviving family member.[13]

So whether a person dies as a result of either brain or cardiopulmonary failure, that person's body enters a new state of being, one in which management of the "dead self" becomes a job for everyone else but the self. The corpse, thus legally recognized as a body with some rights under the law, enters a state of existence where the seemingly normalized rules of scientific, legal, and medical "life" become entirely destabilized. Yet the human corpse is not an entirely passive body upon which these laws and technologies are simply deployed. As Gilligan and Stueve point out, the dead body becomes quasi-property, meaning that a legal guardian or sovereign authority must do *something* with the corpse. In the hours and days after dying, a corpse produces all kinds of results, actions, and scenarios that turn the legal definition of a person into a postmortem "bundle of rights." Within that bundle is a dead human body with a history and theoretical positioning all its own.

* * *

Technologies of the Human Corpse is organized into seven chapters that explore the histories, technologies, political concepts, and human practices that persistently alter the dead body. The first chapter, "Embalmed Vision," presents the history and theory of nineteenth-century preservation technologies that mechanically altered the human corpse.[14] By analyzing the sociohistorical effects of those human technological practices, it is possible to identify how entirely new postmortem conditions for all dead bodies were produced. These technologies of preservation effectively invented the modern corpse, transforming the dead body into something new: a photographic image, a train passenger, a dead body that looked alive. All of these technological innovations are matched by the emergence of an early-twentieth-century funeral industry that turned the preserved human corpse into a dead body that was quasi-atemporal. Once the human corpse could exist outside the normal biological time that controlled the body's decomposition, it became a well-suited subject for unfettered public display. Modern dead bodies emerge as the products of nineteenth-century human technologies that created a kind of embalmed vision that we living humans still use today, often without noticing, when looking at death.

Chapter 2, "The Happy Death Movement," moves through the twentieth century and focuses on death and dying during the 1970s. The years leading up to World War II maintained the nineteenth-century template discussed in chapter 1, but end-of-life practices radically changed during the post-WWII period, and 1970s social movements facilitated those changes. So, for example, many of today's death and dying social movements germinated during the 1970s—for instance, contemporary Death Cafes where people gather to discuss end-of-life issues began forty-five years ago as the Death Awareness movement. A truly robust twenty-first-century death, dying, and dead body historiography must include the 1970s, or what I call the long 1970s moment, e.g., Elisabeth Kübler-Ross publishes *On Death and Dying* in 1969 and AIDS epidemic protests led by the AIDS Coalition to Unleash Power (ACT UP) take 1970s social activism to the streets during the 1980s. One particular 1970s death and dying book that deserves special attention is Lyn Lofland's *The Craft of Dying*, originally published in 1978 and reissued by the MIT Press in May 2019. Chapter 2 offers a close read of Lofland's foundational ideas and why they remain entirely relevant today by presenting the introduction I wrote for the fortieth anniversary edition of *The Craft of Dying*.

A few years after Lofland's book first appeared, unprecedented numbers of gay men and intravenous drug users began dying in the early 1980s from what would later become known as HIV/

AIDS. Chapter 3 focuses on the epidemic's production of "The HIV/AIDS Corpse" and the institutional effects those corpses had on the American funeral service industry. What the HIV/AIDS corpse posed was a direct challenge to the institutional controls developed by funeral directors to normalize and transform the dead body. Many American funeral directors simply did not want to touch a dead body produced by HIV/AIDS, and their anti-AIDS sentiments often made embalming the corpse impossible. How the funeral service industry reacted and changed in response to the emergence of the HIV/AIDS corpse offers an opportunity to reexamine the productive potential of the dead human body. The technologies of the corpse are explicitly used to analyze both these changes to the funeral industry and the productive qualities of the HIV/AIDS corpse. What emerges from the institutional challenges posed by the HIV/AIDS corpse is a specific kind of dead body that has political possibilities for both the concept of a queer politics, as suggested by Catherine Waldby, and the broader subject of human death.[15]

In an October 2006 interview with National Public Radio, *Body Worlds* creator Gunther von Hagens discussed two potential exhibitions.[16] One exhibit would mix human and animal body parts together in order to create mythological creatures. The other would feature male and female corpses having sex. Von Hagens described both of these exhibitions as mere possibilities and gave no further details. Then, in May 2009, von Hagens opened a new exhibition in Berlin, Germany, called *The Cycle of Life*, and, as promised, he posed two different pairs of dead bodies in sexual positions. Each couple consisted of a man and a woman engaged in heterosexual sex. Unfortunately, the May 2009 exhibition did not involve mythological creatures, nor has von Hagens yet attempted to display such creatures.[17] In *Body*

Worlds's over twenty-year existence, it has continually turned the dead body into something new. It is this quest for "newness" that von Hagens is arguably embracing with *The Cycle of Life* exhibition. Von Hagens's exhibitions financially succeed by using anatomical science's history and language to produce popular culture narratives about the dead body. His exhibitions also succeed at offending large numbers of people, and the dead bodies engaged in sex acts proved no exception. Chapter 4, "Plastinating Taxonomies," embraces von Hagens's provocative dead body poses and critically analyzes his work's larger ramifications in terms of human taxonomy, cadaveric anatomy, and death.

Right now, around the world, a seemingly invisible "black market" thrives on tissues, bones, and limbs from dead bodies. Chapter 5 examines "The Global Trade in Death, Dying, and Human Body Parts" and is an analysis of the so-called "underground traffic" in human corpses. It also discusses where the trade in human body parts is headed, relative to the global funeral industry, and the overall necroeconomic value of the human corpse made possible by the American biomedical industry's "body brokers." These are the people who work as middlemen between buyers (which can include biomedical corporations, plastic surgery centers, etc.) and the three primary sources of cadavers: university medical schools, funeral homes/crematoriums, and municipally run morgues that perform autopsies.[18] The endgame logic of the science and technology used on dead bodies since the mid-nineteenth century has always focused on transforming the corpse's flesh into a commodity, a process that turns the human cadaver into a never-ending site of potential productivity. The only real limitation to fully commercializing the human corpse is the onset of decomposition and a lack of imagination with respect to how to maximize the dead body's financial output. Dealers in body parts are also actively helping to produce, knowingly or not, the next logical step for the technologies of the human corpse. This next step is more global in its reach, but no less local in its postmortem biomaterials procurement strategies. This next step also, and importantly, involves the American funeral industry.

From the nineteenth century onward, and in a most discontinuous fashion, postmortem human politics has inserted itself into dead body technologies of all kinds. Giorgio Agamben (borrowing from Michel Foucault) defines two of these key political interfaces, biopolitics and thanatopolitics, as the power to *make live* and the power to *make die*.[19] Chapter 6, "Biopolitics, Thanatopolitics, and Necropolitics," presents a third interface, necropolitics, as the power to *make the human body dead* without the acknowledgement by authorities of any actual "death" transpiring.[20] The human body literally goes from a living state to a dead one without any intermediary steps or considerations. In other words, necropolitics produces dead bodies without death in any recognizable form. Chapter 6, also analyzes and theorizes how a politics of life, death, and the dead body functions in relationship to sovereign power. The sum total of this argument focuses on the fundamental changes taking place in the concept of "the human" as different individuals and groups attempt to control death and the dead body. It is the ever-increasing political transformation of the historically located technologies of the human corpse (e.g., the embalming machine, the postmortem photograph, etc.) into *death prevention technologies* that makes total control over human mortality seemingly viable. But then in chapter 7, "Patenting Death," as Michel Foucault and Giorgio Agamben argue in their own distinct ways, the consequences of total human control over death and the dead body represent not liberation but the total abandonment of *Homo sapiens* as a concept and species.

Most importantly, all of these arguments about new and old technologies surrounding the human corpse, from early embalming machines to future deathless bodies, do not represent a technologically determined model of humanity, but rather a series of possible scenarios. Technologies require tools and practices, so when it comes to dead body technologies, even the most nonindustrial-looking scenario (e.g., natural burial) still requires *humanly invented tools and practices*. We humans use these technologies; the technologies do not use us. Indeed, "blaming the technology" remains a bad habit when discussing death, dying, and dead bodies, so before any critique of postmortem technologies truly begins, we must first look at how we living humans decide to use any given tool. As Raymond Williams suggests in *The Politics of Modernism*, "The moment of any new technology is a moment of choice."[21] Death for the human is quickly becoming one of those choices, with little forethought or consideration as to what living without dead bodies means for the concept of life.

6/12/2018

Watching My Sister Die—City of Lakes/City of Death
I'm back old friend.
In my city of Lakes.
    the place I spent too long before finally leaving
And I find myself discussing nothing but death
which is my job
but now it's personal and it shows.
I've already begun planning the ink
for my sister.
Knowing that she'll soon adorn my body
next to all the other relatives.
Far too soon and before her time.
Also watching my uncle die and the concluding chapter
of my family story taking shape.
You asked me Death if I ever get tired of what I do.
And the answer is no. I can't. This is the only reality
I've ever known.
A reality that I won't abandon now. It's not possible.

And to be honest, it is death that makes me comfortable.
    A city that I can inhabit.

1 Embalmed Vision

I want to say a few words to the American funeral directors and embalmers: Six months ago I made a few gallons of Bisga Embalming Fluid. ... I was finally able to produce a chemical which would not only remove the discolorations, if present, but would combine with the dark discolored blood and restore it to its natural color, thus removing every vestige of discolorations and thus give the body a perfect life-like appearance.

—Carl Lewis Barnes, Bisga Embalming Fluid advertisement, 1902

Dead bodies always attract attention, so in October 1902, Dr. Carl Lewis Barnes and his brother Thornton H. Barnes, both instructors at the Chicago College of Embalming, created a large exhibition of embalmed corpses and body parts for the National Funeral Directors Association in Milwaukee, Wisconsin. The Barnes brothers' exhibit featured human specimens preserved with Bisga Embalming Fluid—a product invented and produced by Dr. Barnes for commercial use by other embalmers. An early trade journal for professional embalmers and funeral directors, appropriately named *The Casket*, printed the following description of the exhibition in its December issue: "Over fifty specimens were on exhibition and the walls of the room were covered in photographs showing the apartments of the Chicago College of Embalming, special drawings of dissections of the arteries, veins and organs. ... No more striking exhibit was ever made than this one. It represented just what Bisga will do when properly used."[1]

The centerpiece of the exhibit was the Bisga Man, an embalmed male corpse sitting upright in a chair, with one leg crossed over the other, and wearing a fashionable suit. According to the story in *The Casket*, the Bisga Man was "one of the most remarkable specimens of embalming ever produced in this country."[2] The emphatic response to the Bisga Man was a testament to how striking it was at the time to see a corpse looking so incredibly alive. In early-twentieth-century America, the Bisga Man represented the perfect nexus of mid-to-late-nineteenth-century preservation technologies that radically redefined the organic existence of the human corpse. These preservation technologies represented a series of overlapping human choices, embalming chemicals, machines, and funeral practices, all intent on keeping the dead body looking "properly human." Yet these industrial forces acting on the human corpse did much more than suspend decomposition by altering the dead body's chemical physiology: through these forces, the concept of human death was itself being simultaneously altered.

It was through this industrialization of the dead body in mid-nineteenth-century America that the modern human corpse also became an invented and manufactured consumer product. More specifically, the modern human corpse became (and remains) an invention of specific mid-nineteenth-century embalming and photographic technologies that seemingly stopped the general public from seeing the visible effects of death. Additional, disparate technologies for handling dead bodies, such as the availability of transcontinental railroad transport and late-nineteenth-century American funeral industry practices, also then merged to expand and control the amount of time available for handling corpses. The merging of these seemingly disparate technologies during the latter part of the nineteenth century not only produced a new understanding of death for twenty-first-century Western nations but, more importantly, they invented a timeless human corpse that resisted organic decomposition and visual degradation. In their own specific ways, each of these technologies helped invent a completely new concept of the dead body.

Postmortem Subjects and Postmortem Conditions

Given the twenty-first-century corpse's organic stability born from nineteenth-century embalming innovations, the day-to-day (albeit largely invisible) presence of dead bodies creates a contradiction: the technologically altered dead body remains a somewhat underexamined subject.[3] Various medical, political, economic, philosophical, and religious discourses include the history of the corpse, but these same discursive groups often overlook the mechanically produced, dead human body.[4] The term *postmortem subject* most accurately defines this overlooked dead body and encompasses the social, political, and biological conditions that surround death, making the ever-malleable human corpse commodifiable.[5] As such, the emergence of the postmortem body—as in a body with a "life" *after life*—is inextricably connected to the nineteenth-century technologies altering the human corpse.

The historical emergence of the postmortem subject can therefore be seen as the moment when preserving the corpse became a human action that made the dead body look more alive. While these technological modifications to the human corpse began in mid-nineteenth-century America, they absolutely continue today in different and often ephemeral forms.[6] Indeed, the *postmortem conditions*, a field of dynamic relations between a human corpse and a living person, consistently illuminate the legal, medical, and biological fluctuations that help articulate both death and the dead body.[7] These postmortem conditions encompass the scenarios of everyday life that legislate and attempt to define an increasingly heterogeneous concept of death for the dead body. As human definitions of death change, so too do the postmortem conditions that produce differing, at times completely oppositional, ontological states. These conditions are also a product of intense, technological changes that increasingly mediate the human body's postmortem state. Nineteenth-century embalming and photographic technologies, for example, became important tools that destabilized and redefined the postmortem state of the corpse by creating entirely new kinds of postmortem conditions.

The human corpse is positioned in these conditions as maintaining a dual functionality as both a deceased subject and a dead object. It is important to use the categories of "deceased" and "dead" with these specific subject/object relations for two reasons. First, the deceased subject is most certainly not a dead object for consumers of medical and funeral practices, such as family members, clergy, and so on. Second, the dead object is not necessarily a deceased subject for users of medical and funeral technologies. This proposed subject/object relation is arguably not a cleanly defined separation where either can be easily located. The ontology of each depends, then, on the observer's relation to the corpse within the fluctuating thresholds of the postmortem conditions. Examining the postmortem conditions produced by nineteenth-century technological changes to the dead body requires using a three-part structure that is both historical and theoretical: the relationship between photography and embalming, postmortem circulation and fluidity, and the construction of a paradoxically "alive but dead" modern, hyperstimulated corpse.

Photography and Embalming[8]

Tom Gunning argues that nineteenth-century American photography allowed the mechanical reproduction of the three-dimensional world and produced "the standardization of imagery" for industrial societies.[9] As the imagery of human life became recorded, printed, and reproduced, so too did human death. Remembering a dead person through photographic evidence of the deceased body emerged as one of the key uses of this new form of visual reproduction.[10] The photographic record of what a dead body looked like began the important process of standardizing how postmortem conditions could or should appear when viewed by the living. To illustrate how popular death photography became in this era, Gunning discusses the production of so-called "spirit photographs" in mid-to-late-nineteenth-century America and Europe. The spirit photographs captured photographic images of a deceased person's soul "miraculously" appearing in the background of family portraits (see figure 1.1). The spirit in the photograph was not limited to family members since "images of Lincoln and Beethoven" sometimes also appeared standing behind the living person being photographed.[11]

Figure 1.1

Spirit Photograph, Man and His Departed Family. Source: Courtesy of Stanley B. Burns, MD, and the Burns Archive. (From *Sleeping Beauty II: Grief, Bereavement and the Family in Memorial Photography*, Burns Archive Press, 2002, p. 52.)

In Gunning's analysis of popular nineteenth-century attitudes about photography, he argues that the public believed that
the machinery of the camera could somehow "see" the world of
death in ways regular human vision could not. Gunning explains
how and why nineteenth-century photography consumers came
to believe in the seemingly impossible appearance of a "spirit":
"... ghosts invisible to the human eye are nonetheless picked up
by the more sensitive capacity of the photograph."[12] Of course
these spirit photographs were doctored images, where the pho-
tographer had superimposed the "ghost" of the deceased person into the image at a later time.[13] The public acceptance of
the spirit photographs suggests, however, that the power of the
machinery producing the images could really capture the "true"
appearance of death as it affected the living. Spirit photography
did not itself become the main form of mourning photography,
but the ability to photograph a deceased person turned human
death into a stable, temporally fixed image, capturing a body
unchanged by organic decomposition. As Jay Ruby explains in
Secure the Shadow, his book on the history of death photography: "Photographs commemorating death can be seen as one
example of the myriad artifacts humans have created and used
in the accommodation of death. Because the object created,
i.e., the photograph, resembles the person lost through death,
it serves as a substitute and a reminder of the loss for the individual mourner and for society."[14]

As the postmortem photograph came to define the appearance of the dead body for years to come, the era's photographers
strove to capture the most vital image possible. Photographs of
the corpse were quickly taken to prevent signs of decomposition from appearing in the image and to capture the deceased
person looking as "alive" as possible. "During the first 40 years

of photography (ca. 1840–1880)," Ruby explains, "professional photographers regularly advertised that they would take 'likenesses of deceased persons.' Advertisements stating that 'We are prepared to take pictures of a deceased person on one hour's notice' were commonplace throughout the United States."[15] The rapid availability of both photographers and photographs meant that the postmortem state of a dead body could be fixed into a reproducible image far faster and with less expense than, for example, death portraits painted on canvas a century earlier.[16]

Ruby goes on to explain that "[t]he custom of photographing corpses, funerals, and mourners is as old as photography itself. It was and is a widespread practice and can be found in all parts of the United States among most social classes and many ethnic groups."[17] The freezing of the corpse's image, "the fascination with death and its overcoming through the technical device of mechanical reproduction,"[18] produced a visual index of *how* death appeared for many Americans during the nineteenth century (see figures 1.2–1.4). A dead person's image could now become a corpse from anywhere at any time, no longer limiting the visual consumption of human death to immediate family. Individual human memories of a dead person's body, in the era before photography, were often confounded by that same body's inevitable state of decomposition, meaning that viewing the corpse needed to occur before it began breaking down.[19]

Through photography, large numbers of people beyond immediate family members could look at a lifelike representation of the dead body indefinitely and in any location the image traveled. As a mechanical apparatus, the camera effectively removed the dead body from any time and space constraints created by death. Images of death suddenly became mobile visual accounts of what a person looked like at the time of burial. The

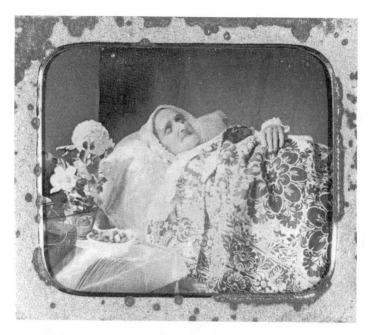

Figure 1.2
Postmortem photograph of unidentified woman. *Source:* Courtesy of the George Eastman Museum (2019). (From *Secure the Shadow*, MIT Press, 1995, p. 68.)

human corpse was also the perfect subject for early photography since it lacked any bodily movement. Southworth and Hawes, a prominent mid-nineteenth-century Boston daguerreotype and photography studio, ran the following advertisement playing on these same public fantasies about vital-looking dead bodies: "We take great pains to have Miniatures of Deceased Persons [wallet or postcard size images] agreeable and satisfactory, and they are often so natural as to seem, even to Artists, in a deep sleep."[20] The visual index mechanically produced by photographers

Figure 1.3
Postmortem photograph of mother with dead child. *Source:* Courtesy of
the George Eastman Museum (2019). (From *Secure the Shadow*, MIT Press,
1995, p. 91.)

Figure 1.4
Postmortem photograph of mourner with dead woman. *Source:* Courtesy of the George Eastman Museum (2019). (From *Secure the Shadow*, MIT Press, 1995, p. 95.)

created a profoundly new way of seeing the dead body for the general public. A letter written in 1870 from Eva Putham to her Aunt Adelaide Dickinson Cleveland describes the photograph taken of Adelaide's dead sister Mabel and offers a vivid example of how strong the mechanical-visual control of the corpse had become: "How lonely you must be, how could you endure it. If it were not for the assurance that it's all for the best. I am glad that you could get so good a picture of the little darling dead Mabel as you did, the fore head and hair look so natural."[21] By the late 1800s, advertising about postmortem photographic services began to disappear from both general circulation periodicals and photography journals since public knowledge of death photographs had become widespread.[22]

Death photography continued unabated until another nineteenth-century technological innovation altered postmortem conditions for human bodies. The new technology was mechanical embalming, and it slowly but surely became a normalized postmortem practice in the United States after the Civil War.[23] Robert Habenstein and William Lamers explain in *The History of American Funeral Directing* that the Civil War was "the first conflict to see embalmers waiting and working in camps, on battlefields, in government hospitals, and in nearby railroad centers, to serve the needs of the military and the families of the fallen."[24] After the Civil War ended in 1865, the practice of embalming had become so widespread for Union Army soldiers killed during battle that embalmers had a new consumer product to offer the general public: the mechanically preserved dead body.[25] The photographic images of dead persons offered by firms such as Southworth and Hawes defined and standardized a history of postmortem visual culture that played an integral part in the production of lifelike corpses for late-nineteenth-century embalmers. The use of photography did not completely disappear from the early-twentieth-century funeral industry; rather, embalming allowed the actual body to be on display without the requirement of a fast burial. Instead of eliminating postmortem photography, embalmers simply copied the photographic image's aesthetic (to produce the effect of "deep sleep") for the dead body.

The nineteenth-century embalming process involved removing the bodily fluids of a corpse through a handheld vacuum pump. Once the organic fluids were removed, the pump was used to reinject the dead body with a preservative chemical solution (see figures 1.5–1.8). Mechanical embalming's fundamental

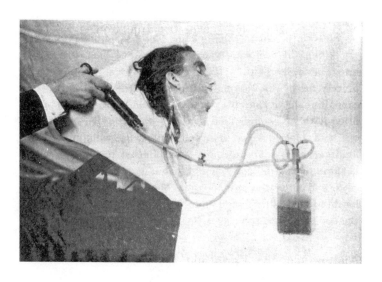

LINEAR GUIDE.
Needle in position. Operator injecting the body.

Figure 1.5
Embalmer injecting the corpse with fluid. *Source:* Courtesy of the National Funeral Directors Association (USA). (From *The Art and Science of Embalming*, Trade Periodical Company, 1896, p. 248.)

historical importance was that it marked the radical slowing of decomposition by overriding the human body's biology. This mechanical-chemical process ultimately produced a dead body that remained unaltered by time.

The embalming process produced rather "amazing" effects, as an advertisement from an 1863 Washington, DC, business directory proclaimed:

Bodies Embalmed by Us
NEVER TURN BLACK!
But retain their natural color and appearance; indeed, the method having the power of preserving bodies, with all their parts, both internal and external.

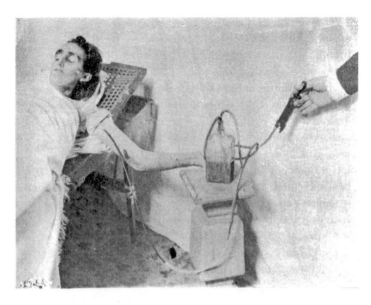

OPERATOR REMOVING BLOOD FROM THE BODY; FLEXIBLE SILK TUBE INSERTED INTO THE
LEFT BASILIC VEIN AND ENTERING THE RIGHT AURICLE OF THE HEART.

Figure 1.6
Embalmer removing blood from corpse. *Source:* Courtesy of the National
Funeral Directors Association (USA). (From *The Art and Science of Embalm-
ing*, Trade Periodical Company, 1896, p. 141.)

WITHOUT ANY MUTILATION OR EXTRACTION,
and so as to admit of contemplation of the person Embalmed, with
the countenance of one asleep.[26]

The Art and Science of Embalming (1896), a nineteenth-century
textbook written by Dr. Carl Lewis Barnes, goes to great lengths
to explain the power represented by properly embalmed bodies.
Barnes provided four key reasons for embalming the dead body:
[1] "to prevent the appearance of putrefaction until such time
that the body may be viewed by the friends of the deceased, or

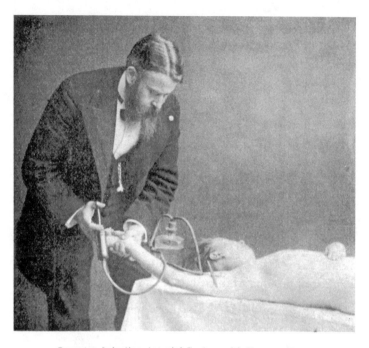

Operator Injecting Arterial System with Vacuum Pump.

Figure 1.7
Embalmer injecting fluid into corpse. *Source:* Courtesy of the National Funeral Directors Association (USA). (From *The Art and Science of Embalming*, Trade Periodical Company, 1896, p. 231.)

until it can be conveyed to a suitable resting place ... *[2]* that of disinfecting the body ... *[3]* preservation of a body until it may be identified. ... *[4]* A fanciful reason for embalming is that the body is made to look life-like."[27] Barnes clearly explains, point by point, that the power of embalming is total control of the dead body, to prevent any kind of disease or organic breakdown and to maintain the body's aesthetic appeal. Later, in the same

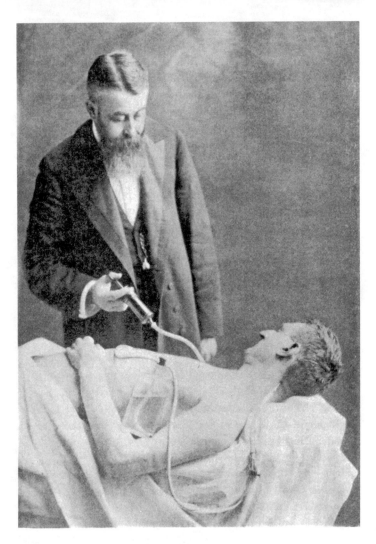

The Barnes Needle Process.

Figure 1.8

Embalmer using the Barnes embalming method. *Source:* Courtesy of the National Funeral Directors Association (USA). (From *The Art and Science of Embalming*, Trade Periodical Company, 1896, p. 251.)

textbook, Barnes likens the work of the embalmer to assisting the work of God: "EMBALMING prevents the corruption of the grave, so that the body will remain entire, and as it were asleep in its bed, till awakened by the last trumpet to a joyful resurrection, where in its flesh it shall see God. ... Hereby death has no more power over us than a long sleep."[28]

The connections that Barnes makes between the total control of the dead body and the Providential virtues of the embalmer's work are significant for both the postmortem condition and visual control of the dead body.[29] Embalming, in a way similar to death photography, affected not only how the viewer observed the corpse's postmortem state but also radically altered the process of gazing at a dead body. This is the crucially important meeting point of photography and embalming: What embalming and death photography fundamentally changed was how the living observer viewed the postmortem conditions of the corpse. In effect, these innovations in embalming practices enabled the emergence of *embalmed vision*. Visually experiencing the corpse for an extended period of time no longer required photographic lenses, since the public could now *see* the dead body by using the pumps and needles of the Barnes embalming method. This crucial shift for the nascent American funeral industry standardized an experience of the corpse that persists today, and it altered the popular understanding of how "natural" or "normal" death appeared. In his history of death in America, James Farrell cites two examples of this early visual control over the dead body: "W.P. Hohenschuh [an Iowa funeral director] advised that 'one idea should always be kept in mind, and that is to lay out the body so that there will be as little suggestion of death as possible.'" A few years later in 1920, "a Boston undertaker allegedly advertised: For composing features, $1. For giving the features a look of quiet resignation, $2. For giving the features the appearance of Christian hope and contentment, $5."[30]

As the prices listed by the funeral director in Boston suggest, the embalming process and the embalmed vision it produced meant that the corpse could now be commodified in new ways. Since the tools used to alter the corpse were largely invisible and the embalming process took place behind closed doors, the end result was a new kind of postmortem subject that looked alive, yet bore no visible marks of mechanical intervention. The embalming tools' sheer invisibility meant that the mechanical apparatus could itself become permanently attached to a viewer's understanding of the dead body, without directly altering the observer's conscious understanding of death. Almost imperceptibly, the act of looking at a corpse in the late nineteenth century shifted from potentially seeing uncontrolled decay to seeing a mediated image of the dead body. As the previously unregulated postmortem conditions of the dead body came under control, first through photography and later through embalming, the living observer's embalmed vision effectively cocooned the corpse. The dead body was placed in a space of death stripped of any adverse smells, appearances, diseases, or even human mortality itself. The corpse was no longer controlled by biological death in the late nineteenth century; rather, control of the corpse and the "death" that it presented shifted to human actors.

As soon as the dead body began to look less dead, the need to reproduce and consistently reaffirm embalmed vision (over time and across geographical boundaries) created an entirely new market for handling human corpses. The relationship between photography and embalming took the decomposition of corpses and turned it into an economically profitable business. While the embalming process helped transform the human corpse into a new consumer product, it also served to eliminate the perceived public health disease threats associated with circulating the dead in the public sphere by making the deceased body "safer." Embalming's hygienic popularity resulted in the growth and marketing of many fluid-based, preservative disinfectants, but it also led to a number of problems for both the emerging early-twentieth-century funeral industry and for dead bodies.

The Politics of Circulation and Fluidity

The terms *circulation* and *fluidity* have dual meanings for the technologically altered human corpse, but both sets of meanings represent important aspects of what happened to dead bodies, in America, during the nineteenth century. Circulation and fluidity are used here to mean both the movement of fluids through the human body and the railway shipment of dead bodies circulating across the United States. The use of these terms is partially rhetorical, but more accurately material. Each term serves as a means to articulate how the nineteenth-century human corpse became the nexus for controlling both the temporality of death via embalming and the space of death with train travel. Shipping the dead body on trains added yet another important level of human technological control over the dead body.

The history of how dead bodies could be made safe for transcontinental travel in America is a story closely tied to the mergers of chemical embalming and the emergence of city-to-city rail transit. The American population's expansion across the continent meant that corpses needed to travel the same distances and routes. Farrell comments, "During the Civil War, some people embalmed bodies for shipment home from the front. After the war, embalmers continued to prepare the bodies of formerly

mobile Americans for shipment to family and friends."[31] Habenstein and Lamers highlight the multiple issues encountered when
shipping dead bodies during the post–Civil War nineteenth
century, noting that "it became necessary to inspect trains at
depots for improperly embalmed bodies, broken shipping cases,
bodies bearing the germs of infectious diseases and the like."[32]
They go on to explain how "exasperated" funeral directors
from across the country often described "the chaos of uncodi-
fied and non-articulatory rules and regulations dealing with this
problem."[33]

Funeral directors and train companies that shipped bodies,
both immediately following the Civil War and into the early
twentieth century, faced a serious situation: the propensity for
substandard embalming jobs done by poorly trained embalmers. In October 1907, the 4th Annual Joint Conference of the
Embalmers' Examining Board of North America convened in
Norfolk, Virginia, to discuss the problems arising from inad-
equately embalmed bodies being shipped across the United
States. The Embalmers' Examining Board of North America's
president, Dr. H. M. Bracken of St. Paul, Minnesota, delivered
the conference's annual address and explained the need for individuals preparing the dead for transport to receive a more thorough and standardized embalming education: "So long as the
undertaker was interested only in preparing the remains of the
dead for immediate and local burial he had a simple task before
him. But with the age of travel and migration a new responsibility was thrust upon him, viz.: the preservation of a body
in order to permit of its shipment. This responsibility changed
the undertaker to the embalmer."[34] Training individuals in how
to properly embalm dead bodies for rail shipment, however,
also created opportunities for unscrupulous con men. Bracken

explains, "As the quiz master came into existence to prepare the medical student for his examination, so the 'fluid Man' sprang up to provide embalming fluids and embalmers' quiz classes for the embalmer."[35]

Bracken's description of the "fluid Man" was a term used to describe individuals who worked as itinerant embalming educators, opening "schools" across the country to teach would-be embalmers how to preserve dead bodies. The instructors were themselves mostly unqualified to teach embalming, and their work often yielded dubious results.[36] Habenstein and Lamers comment: "It is interesting also to note that the earliest compounders of embalming fluids, whether medically trained or not, chose to call themselves 'professors... .' Thus a good decade before the appearance of any type of formal instruction, 'Professor' E. Crane had patented in 1868 and sold 'Crane's Electro-Dynamic Mummifier.'"[37]

In 1906, a year before Bracken's address, the professionals most affected by the work of the "fluid Man" had already convened a special meeting to establish and standardize railway transport rules for the dead body.[38] The meeting's main participants included the National General Baggage Agents' Association, the National Conference of Health Officers, and the National Funeral Directors Association. The "Transportation Rules" approved at the meeting stipulated a series of nine, self-imposed regulatory mandates that directly addressed the problems created by bodies improperly prepared for shipment.[39] The rules covered an array of diseases (yellow fever, anthrax, etc.) as well as how the dead body should be shipped.

Rule 1 was the most explicit: "The transportation of bodies dead of small pox and bubonic plague, from one state, territory, district or province to another, is absolutely forbidden."

Rule 2 required a state-certified embalmer to properly prepare any dead body exposed to disease by embalming it, disinfecting it, and putting "absorbent cotton" in any orifice. After completing those procedures, "such bodies shall be enveloped in a layer of dry cotton not less than one inch thick [see figure 1.9], completely wrapped in a sheet securely fastened and encased in an air-tight zinc, copper or lead lined coffin, or iron casket, all joints and seams hermetically sealed, and all enclosed in a strong, tight wooden box" [see figure 1.10]. After the corpse was properly prepared for shipment, Rule 6 required that "every dead body must be accompanied by a person in charge, who must be provided with a passage ticket and also present a full first-class ticket marked 'corpse.'"[40] The full nine transportation rules cover a variety of other logistical issues, including guidelines for the dead body's "express transit." Guaranteeing the safe passage and proper delivery of a dead body meant creating a whole new regulatory regime for the emerging nineteenth-century

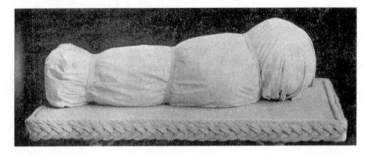

Body ready for transportation.

Figure 1.9
Corpse wrapped for train transport. *Source:* Courtesy of the National Funeral Directors Association (USA). (From *The Art and Science of Embalming,* Trade Periodical Company, 1896, p. 349.)

Box ready for transportation to any part of the world.

Figure 1.10
Corpse shipping casket. *Source:* Courtesy of the National Funeral Directors Association (USA). (From *The Art and Science of Embalming*, Trade Periodical Company, 1896, p. 350.)

transportation industry. Mechanical embalming was a crucial part of that new regime since it enabled chemical control over the dead body's circulation and fluidity.

In addition to possessing a first-class railway ticket, an embalmed corpse could now also circulate outside conventional time and space through these same preservation methods and transport systems. As a general rule, the space and time occupied by living bodies and dead bodies functioned differently. Yet with the advent of nineteenth-century embalming technologies, the living could usurp this temporally adjusted *corpse time* and make dead bodies conform to rules not within the human gene code, that is, the automatic postmortem onset of organic decomposition.[41] Wolfgang Schivelbusch most accurately describes these kinds of temporal changes in *The Railway Journey: The*

Industrialization of Time and Space in the 19th Century. While Schivelbusch does not directly address train travel's effects on dead bodies, his overall analysis of how a radically altered time code affected the circulation of living bodies absolutely includes the human corpse. Schivelbusch argues: "'Annihilation of space and time' was the early nineteenth century characterization of the effect of railroad travel. The concept was based on the speed the new means of transport was able to achieve. ... In terms of transport economics, this meant a shrinking of space."[42]

The unembalmed dead body resisted this new temporal annihilation better than many other human bodies at the time. This meant that the total preservation of the human corpse became a necessity in order to make the dead body controllable for both train travel and transport economics. The shipping problems encountered by funeral directors, baggage workers, and state health officials help illustrate Schivelbusch's additional point that "the alteration of spatial relationships by the speed of the railway train was not simply a process that diminished space, but that it was a dual one: space was both diminished and expanded."[43] By the early twentieth century, regulated embalming similarly diminished decomposition and made expanded travel by most dead bodies possible—setting aside cases of infectious disease such as bubonic plague. The enlarged distance required to transport a dead body no longer posed a problem since the entire concept of corpse time and cadaveric decomposition had largely been slowed.

Two very different kinds of nineteenth-century machines, embalming tools and trains, subsequently merged to exert control over the human corpse. This industrialized control of the dead body meant that embalmed vision could function without concern for the space and time of death. Once embalming slowed conventional time from affecting the dead, the organic space of death became replaced by something new for the modern corpse—a mechanically produced space of ontology. The control exerted over the human corpse in the nineteenth century also meant that American funeral directors suddenly acquired a great deal more power as both licensed embalmers and producers of the dead body's final appearance. Just as properly embalmed bodies helped the railway companies move corpses across state lines, the transportation rules also helped funeral directors further legitimate their work. Habenstein and Lamers explain the long-term benefits of the new shipping regulations for funeral directors: "Finally, they gained *professional* recognition for their work in embalming; that is, the affidavit of a funeral director that he had embalmed a body was accepted at face value henceforth by baggage agents and public health functionaries alike."[44] Early-nineteenth-century postmortem conditions and corpse time were spaces initially outside concrete human control, but control was finally exerted as new technological methods normalized the "healthy" looking dead body. What the dead body's circulation and fluidity on trains quietly created for American funeral directors was the total consumption of postmortem space and time.

Construction of the Modern, Hyperstimulated Corpse

The unveiling of the Bisga Man by Dr. Carl Lewis Barnes in 1902 marks a particularly historic moment for the modern corpse.[45] The Bisga Man's creation introduced a highly compelling paradox: the invention and marketing of the ultramodern and hyperstimulated human corpse (see figure 1.11).[46] Barnes used the Bisga Man to literally and figuratively embody the catalog

Figure 1.11
Bisga Fluid advertisement featuring the Bisga Man. *Source:* Courtesy of the National Funeral Directors Association (USA). (From *The Sunnyside*, 1902, National Funeral Directors Association archive.)

of changes made possible by nineteenth-century preservation technologies. The Bisga Man was also the perfect human model for the most powerful promise made by nineteenth-century embalming and preservation: dead bodies could suddenly appear unnaturally alive.

From the Civil War era onward, the ontology of the corpse becomes a highly productive relationship for living and dead bodies alike, and represents not a reversal but rather an extension of the space occupied by the dead body. The Bisga Man is an example of a corpse living within that new postmortem ontological structure, demonstrating how postmortem conditions also impinged upon the existential questions raised by mechanically altering dead bodies. The photograph used in the Bisga embalming fluid advertisement is an illustration of a human corpse doubly altered by nineteenth-century preservation machines: first by embalming and then by photography. In a break from nineteenth-century death photography norms, the Bisga Man's image was not a postmortem reminder of the deceased person. Instead, his entire existence as a marketing tool was made possible by merging the power of photography and embalming. The caption underneath the Bisga Man photograph encapsulates that new power by simply stating "Picture taken three months after death." Another version of the advertisement testified that Bisga Fluid produced a "perfect life-like appearance."[47] The amount of time that had elapsed since the Bisga Man's death was crucial in demonstrating to other embalming practitioners just how alive they could make their own dead bodies look if they opted to technologically control the visual experience of human death.[48]

Barnes's advertisement is especially compelling since it manipulated the postmortem conditions of embalmed vision in a counterintuitive way: without the caption describing the Bisga Man as "after death," the body actually looked alive. The Bisga Man exists as a modern corpse whose life and death is impossible to recognize or decipher without the assistance of the caption. Death, as represented by the Bisga Man's postmortem vitality, is reduced, on the one hand, to a textual prompt and is expanded by Barnes, on the other, into a marketing strategy. The alive-looking dead body's technical production through mechanical means, and the subsequent suspension of death from the corpse, meant that modern technologists had found a way to make human death mechanically disappear. From the early-twentieth-century viewpoint of embalmers and funeral industry workers, this new kind of corpse was an object of vast capital potential that simultaneously remained a subject of loss for mourners. With the emergence of machine-based embalming systems, and to a certain extent photography, the modern corpse could suddenly function longer and be more profitable for all living parties. Dead bodies that resist decomposition are perfect candidates for a whole range of postmortem applications, and these conditions make death highly profitable for the industries turning the dead body into an unfettered source of capital.

Death and the dead body in America were radically altered by nineteenth-century technological control over the corpse's inner chemistry and outer appearance. The modern corpse's invention was a product of various invisible human machines, infiltrating and hyperstimulating the dead body for the multiple and varied purposes of the living. What made the development and use of embalming so important was the absolute invisibility through which it functioned. The process of embalming was hidden from view, as were the chemicals used by embalmers. What emerged from nineteenth-century mechanical labor on the human corpse was a modern dead body that required embalmed vision to be seen in the proper context and state of vitality. The corpse that emerged in the late nineteenth and early twentieth centuries persists today in its capacity to remain both lifelike and visible. If left unmediated, the dead body is a shocking spectacle for those not accustomed to seeing physical decomposition. Death photography, mechanical embalming, the dual meanings of fluidity and circulation for dead bodies, and the invention of modern corpse ontology were all used to produce late-nineteenth-century postmortem conditions. Most significantly, they each served to condition a dead body that would undergo further, radical changes in the years to come.

7/29/2018

Watching My Sister Die—#21. Julie Post
I'm on the plane right now
little sister
Flying to you and your dead body
You died 30 minutes ago
Your body finally gave up after you pushed so far.
I said goodbye but I still wanted to be there
today
when you died.
If only you could have waited five-more hours ...
But I know you waited long enough
None of this was supposed to happen this way.
You were never supposed to die first.

But you did.
And tonight when I hold your hand I'll tell you about
all the things people said about you
our friends
After I finally crossed 21. Julie Post off my to do List.
After I held your hand and told you that you were dying.
through tears and tears and you telling me → I'd do the same for
        you.
And you would.   I know it.
It's so beautiful up here, little sister.
10,000 feet above the world
Pale Blue skies and remnants of clouds   drifting past the
British landscape you always enjoyed.
    Laughing about how my job never really made sense
that only your marching to a different   beat brother could
end up as the Overlord of Death
But here's the thing, little sister.
This impossible career only really made sense to me 30 minutes ago
when you died
When I held your hand telling you that you were dying.

When Death warned me that I wasn't prepared for this moment.
Oh little sister—I never wanted to be an only child
you were the one who was supposed to care for Mom and Dad
And now it's all on me, in these final years.

The beginning of one less person in these family photographs.
It doesn't seem possible. All of this. Happening Now. But it is.
The Death Family is not protected or immune from   mortal
          ends.
I'll be   with you soon and hold your lifeless hand, so full of
          love
for everyone around you
And so missed now by the same people.
Gate 11. Bristol Airport. A phone call from our mutual college
          friend.
This is how I learned you died.
So no more watching you die, little sister.
No more pain.
Just sadness and tears and the knowledge that if our roles were
reversed.
You would   write these same exact words for me.

2   The Happy Death Movement

Death and dying during the 1970s is, for me, the crucial decade that everyone should study. The years leading up to World War II largely followed the late-nineteenth-century template discussed in chapter 1, but end-of-life practices radically changed during the post-WWII period, and 1970s social movements facilitated those changes. My scholarly obsession with the 1970s is usually met with confused stares, but it really was the decade when so many of today's death and dying social movements took shape. For example, contemporary Death Cafes where people gather to discuss end-of-life issues was something people did forty-five years ago under the Death Awareness umbrella. Few forms of activism emerge from nowhere, and today's death and dying activists maintain strong ties with often forgotten progenitors. Ironically enough, death and dying social movements during the 1970s were painstakingly documented and described by both journalists and academics. What I find more remarkable is how little of that body of work is remembered today. This collective death activism amnesia also makes complete sense, since every generation reclaims end-of-life issues for itself and turns dying into something new.

But a truly robust twenty-first-century death historiography must include the 1970s, or what I call the long 1970s moment. The long moment begins in the late 1960s and extends into the mid-1980s, as different death and dying events take shape. So, for example, Elisabeth Kübler-Ross publishes *On Death and Dying* in 1969, and 1980s AIDS epidemic protests led by ACT UP and other groups take 1970s social activism to the streets. Even writing that sentence unfairly streamlines over ten years' worth of disparate groups, thinkers, activists, and political agendas. This is also why death and dying in the 1970s deserves its own book-length discussion. Reducing what happened during that decade to even five thousand words is deeply unfair to the era's long forgotten end-of-life advocates.

One particular 1970s death and dying book that deserves special attention is Lyn Lofland's *The Craft of Dying*, published in 1978. Most 1970s death and dying books are long out of print, and Lofland's *The Craft of Dying* was part of this lost group until I convinced the MIT Press to reissue it in 2019. I first read *The Craft of Dying* in 2014 and tried for years to find a way to republish it. Lyn Lofland is an emerita professor at the University of California, Davis, but all my attempts to reach her failed. Then, miraculously, one of her last PhD students, Ara Francis, attended the June 2018 Centre for Death and Society conference on the *Politics of Death*. Ara is an associate professor in the Department of Sociology and Anthropology at College of the Holy Cross in Massachusetts and director of Gender, Sexuality, and Women's Studies. We knew nothing about each other before the 2018 Centre for Death and Society conference, least of all any mutually shared interest in Lyn Lofland and *The Craft of Dying*. In fact, we sat near each other during the conference dinner for a couple of hours discussing student issues, university politics, and academics with tattoos (as one does at these events) without once discussing Lyn. It was only after we were both standing and ready to leave the restaurant that somehow one of us mentioned *The Craft of Dying* and Lyn Lofland. I do not exactly remember who said what first, but I do recall saying that I really wanted to republish *The Craft of Dying* if Ara thought she could get Lyn on board. Ara agreed, and we were off and running.

Reissuing books can sometimes take an enormous amount of work to simply determine who even owns the rights, but *The Craft of Dying* quickly came together. Ara arranged everything with Lyn, and I sold my editor on convincing the MIT Press that a forty-year-old book on death and dying would definitely appeal to today's audiences. What Lyn Lofland might call the twenty-first-century Happy Death Movement.

I wrote a new introduction and Ara wrote an epilogue for the fortieth anniversary edition, both of us reflecting on what *The Craft of Dying* presciently understood about our own death and dying moment. I asked my editor if I could run a slightly edited version of that new introduction in my own book, since it gives a broad overview of the 1970s and I hope encourages as many people as possible to read the reissued *The Craft of Dying*. Ara's epilogue is another important reason to read the fortieth anniversary edition, and I strongly encourage anyone following today's end-of-life politics to pay close attention to her discussion around contemporary hospice care.

Part of me hopes that forty years from now a new generation of death and dying academics discovers my own book and somehow manages to reissue it. I have my sincere doubts that this will happen, but I do think people will still be reading *The Craft of Dying*. It is that important.

Introduction to the 40th Anniversary Edition of *The Craft of* *Dying: The Modern Face of Death* **by Lyn Lofland**

John Troyer

Lyn Lofland's *The Craft of Dying* (1978) is one of the most important books on post-WWII death and dying practices that almost
no one has read. To see Lofland's largely overlooked, but still
relevant, text republished by the MIT Press is both thrilling and
deeply gratifying. It is the one book that in my capacity as director of the Centre for Death and Society at the University of Bath
I think every person working on contemporary death and dying
issues must read. Indeed, I strongly recommend that anyone
interested in understanding how events forty years ago shaped
what Lofland would call today's "thanatological chic" read
The Craft of Dying and note the current uncanny resemblances
to the 1970s.

  The Craft of Dying is, for me, *that* death, dying, and end-of-life
issues book.

  A common response to my adamant recommendation is—
why? Why and how is this specific book any different or better
than its contemporaries, e.g., *On Death and Dying* by Elisabeth
Kübler-Ross or *The Denial of Death* by Ernest Becker (to name two
big death canon contenders)? My rapid answer is that Lofland's
book documents what happened in the 1970s (the formation of
new hospice spaces, activist groups encouraging people to accept
death, the introduction of college courses on dying, and so on)
alongside an invaluable critique of those activities. In fact, it is
Lofland's critique and classification of death-focused groups as
social movements creatively constructing a new end-of-life ideology that makes *The Craft of Dying* fundamentally important.

Lofland calls these end-of-life groups (similar in structure, she
will note, to diffuse 1970s women's movement and environmen-
tal movement groups) the Happy Death Movement and uses the
term to connote enthusiastic warriors taking on a challenge. Her
critique is both generous and insightful at all times. But Lofland
was not content with merely documenting what these death and
dying groups did; she wanted to better understand what moti-
vated their new end-of-life politics and thinking. It is her push
to clearly articulate what is happening in her own moment that
makes *The Craft of Dying* so valuable today; almost every argument and observation she first presented forty years ago remain
both pertinent and urgently needed now.

Her book is truly a message in a bottle, and one sent from
a decade when death and dying social movements coalesced
around end-of-life ideologies that the Western world still struggles with today. That Lyn Lofland accomplished this feat in so
few pages is an achievement in and of itself.

Discovering *The Craft of Dying*

For all my praise of Lofland's work, I am embarrassed to say that
I first learned of, and then read, *The Craft of Dying* in summer
2014. My midcareer discovery of Lofland occurred only after
my esteemed colleague (and walking Death Studies encyclope-
dia) Tony Walter asked if I knew her book and the Happy Death
Movement argument. I said that no, I didn't. Tony asked about
Lofland, because he understood how *The Craft of Dying* directly
related to my ongoing research on American death and dying
discourse during the 1970s.

In a nutshell, this research project examines how the
1970s functioned as a crucial but largely forgotten decade for

understanding what motivates today's death and dying groups, as well as foreshadowing many current end-of-life debates. It is during the 1970s that new death and dying tools and technologies took root, altering the definition of death: things taken for granted today, such as living wills and life-support technologies. Much of the decade's activity was at a very local level and included individuals forming groups emphasizing Death Acceptance, the Right-to-Die, and dying a Natural Death—all thoroughly documented in *The Craft of Dying*.

But the 1970s was also a decade when end-of-life issues extended all the way to the White House and bookended politically tumultuous times. In 1971 President Richard Nixon announced his War on Cancer, and in 1979 President Jimmy Carter formed the President's Commission for the Study of Ethical Problems in Medicine and Biomedical Behavioral Research, which later published its landmark 1981 report *Defining Death: A Report on the Medical, Legal, and Ethical Issues in the Determination of Death* during the Reagan administration.[1] Carter's group would eventually become known as the President's Council on Bioethics and advise future presidents on a wide array of issues, including, but not limited to, death and dying.[2]

Lofland's research remains a key historical and conceptual anchor for anyone interested in that decade, since *The Craft of Dying* analyzed and critiqued what was happening in the 1970s *during the 1970s*. What any reader comes away with from her book, I think, is how death and dying were national conversations related to ongoing events, for example, the Karen Ann Quinlan right-to-die case in New Jersey (which also went global)—and connected to personal freedoms—e.g., the country's first Natural Death Act, passed in California in 1976, and gave individuals the right to legally refuse medical treatments even if the refusal meant dying.

After Tony Walter's helpful nudge, I discovered that *The Craft of Dying* was long out of print (the republishing idea first occurred in this very moment), but I persisted in locating a copy and subsequently devoured the book in one August 2014 sitting. I say in all seriousness that reading *The Craft of Dying* fundamentally changed how I approached all research on death, dying, the dead body, end-of-life concerns, the politics of death, the historical formation of hospice spaces, current Happy Death groups pushing for what Lofland has called "death talk," and neoliberal economic "choice" about funerals. I could go on and on. And like any convert with a newly discovered evangelical zeal, I wanted nothing more than to excitedly read long sections of *The Craft of Dying* to audiences.

Coincidentally enough, captive audiences were available to me in August 2014, since I was the Scholar in Residence at the Morbid Anatomy Museum in Brooklyn, New York (now sadly closed). I am not kidding when I say that almost all my public lectures during the residency involved my simply reading sections from *The Craft of Dying*, especially the introduction:

It seems likely that eventually humans will construct for themselves a new, or at least altered, death culture and organization—a new "craft of dying"—better able to contain the new experience. I believe, as do other sociological observers ... that in the ferment of activity relative to death and dying during the last two decades in the United States we have witnessed and are witnessing just such a reconstruction. Undoubtedly within this ferment, especially that emanating from the mass media, there are elements of fad and fashion—a thanatological "chic" as it were, having approximately the same level of import as organic gardening and home canning among the rich. And certainly one can never underestimate the capacity of American public discourse to transform "life and death matters" into passing enthusiasms. But there is, I believe, more to this activity than simply one more example of impermanent trendiness in modern life. Americans, especially affluent middle-class Americans, have been in the process of creating new or at least altered ways of thinking, believing, feeling, and acting about death and dying because they have been confronting a new "face of death."[3]

And if you are reading this now and thinking to yourself that these words eerily describe death and dying in your own historical moment ("fad and fashion" always gives me pause), then you begin to see why a book published in 1978 continues challenging everyone to examine how any decade's Happy Death Movements can possibly be unique, or new, or revolutionary. Lofland wants readers to understand the history of the present, so that the next generation's death and dying activists might also comprehend the historical relationships to their own current struggle.

Relevance for Today

The Craft of Dying also productively intervenes in one of the 1970s' most unhelpful and unnecessary death and dying arguments, an argument that dogmatically persists today—i.e., that death is a taboo. If the Happy Death Movement functioned like a true social movement, Lyn Lofland reasoned, then that movement needed an enemy, and the death taboo is the perfect foil, since everybody already "knows" that it "exists."

Lofland is neither the first author, nor will she be the last, to thoroughly challenge how and why the death taboo argument is used, abused, and greatly exaggerated. The death taboo will always be a productive fiction for various Happy Death Movement groups committed to ideologically transforming the "face of death" in America and the West, but it is a fiction all the same. As she clearly explains, the death taboo argument serves a useful function that is especially popular with death-movement intellectuals (full disclosure: I am a card-carrying member of said group). Her critique of death-movement intellectuals is reason enough to appreciate how farsighted this text remains today. Lofland's crucial intervention begins:

It has been variously formulated, but essentially the view holds that America is a death-denying society, that death is a taboo topic, that death makes Americans uncomfortable so they run from it, that death is hidden in America because Americans deny it, and so forth. The consequences of all this denial and repression are asserted to be quite terrible: exorbitant funeral costs and barbaric funeral practices, inhumane handling of dying in hospitals, ostracism of the dying from the living, inauthentic communication with the fatally ill, an unrealistic, mechanical, non-organic view of life, and so forth. ... As many scholars have pointed out, the empirical evidence for all these assertions is something less than overwhelming (see, for example, Dumont and Foss, 1972; Donaldson, 1972).[4] And one might consider it somewhat odd that the statement that death is a taboo topic in America should continue to be asserted in the face of nearly a decade of non-stop talking on the subject. But if one appreciates the functions these statements serve in enemy evocation, one can also appreciate that their questionable empirical basis is hardly a serious obstacle to endless repetition. The importance of the "conventional view of death"—of the conventional wisdom about death—as propounded over and over again by movement intellectuals, is not its "truth" but its utility.[5]

If making more people rigorously question whether or not they really need the death taboo fiction to advance their own death and dying arguments is the only thing republishing *The Craft of Dying* accomplishes, then all the waiting was worth it. Seeing the taboo argument finally debunked would also recognize Lofland's scholarly commitment to status quo challenging scholarship both then and now. That said, I have a strong hunch that in the decades to come many death-movement intellectuals and practitioners will still make the death taboo argument to advance both their careers and book sales—a point not lost on Lofland when she states that the death taboo is always about utility, not truth.

Above and beyond the book's uncanny timeliness, Lofland taps into another core human experience: we *Homo sapiens* persist at dying. The fact that we all eventually die becomes that rare universal constant that allows each human the opportunity, should we take it, to experience and think about death and dying in new ways. Lofland focuses, in particular, on how the dying person becomes something different during the 1970s.

I found myself directly confronting Lofland's newly articulated 1970s experience of death and dying when my younger sister, Julie Troyer, died from terminal brain cancer on July 29, 2018. Watching my sister die made me reflect quite heavily on *The Craft of Dying's* key assertions, and in very unexpected ways that accidentally (albeit sadly) coincided with writing this introduction. The MIT Press expressed interest in republishing *The Craft of Dying* while my sister was dying, but I started writing the introduction after she was dead—an interval of approximately one month. My father, Ron Troyer, a longtime grief and bereavement support group facilitator and retired American Funeral Director, best summed up my death interval experience in very Loflandian language: it is one thing to publically say, "Julie is dying"; it's an entirely different experience to state, "Julie is dead." The former felt active, the latter inert.

I chose to add this section about my experience with death and dying, since Lofland rigorously analyzes the role of language and expressivity in encouraging people to discuss precisely these issues. For many days I wondered aloud if it was appropriate for a death studies academic, such as myself, to write a new introduction for *The Craft of Dying* that included a discussion of such a personal experience. After staring at her book for what seemed like eons, I fully realized the genius of Lyn Lofland's irreplaceable contribution to contemporary death and dying discourse: that, no matter what any of us do; no matter our personal, professional, or familial relationship with death, everyone still dies. And that Lofland's always-new-craft-of-dying requires we living humans to critically reflect on these confrontations with mortality in our own meaningful ways, so that we might glimpse, for a moment, what living and dying can become in our technologically advanced twenty-first century. It is vitally important, I think Lofland would say, to see our personal mortal ends in the modern face of death.

What, Then, for the Future of *The Craft of Dying*?

I see no reason why Lofland's book will not remain relevant for another forty years. In surveying how *The Craft of Dying*'s central arguments evolved over time, connections clearly emerge with the ACT UP AIDS protests of the 1980s and 1990s, and the contemporary activism of today's Black Lives Matter groups. Lofland rightly predicts that death and dying social movements will persist at emerging and folding back into each other, precisely because death refuses to phenomenologically disappear. The complexity of what she wrote has never dissipated and will continue to find new audiences for many years to come.

The Craft of Dying does come with a cautionary note, however, and it is a point that bears mentioning in the conclusion to this new introduction.

Happy Death Movement groups (then and now) always run the risk of alienating the very people they so eagerly want to help through nonstop ultra-upbeat expressive death talking that then demands transforming and accepting death/dying/mortality at all costs. The challenge here involves individuals becoming convinced that they are doing death wrong, and in that moment of doubt, Lofland wryly suggests, a "dismal death" movement might emerge:

> If expressivity comes to be widely accepted as the only way to achieve a decent death, the emotionally reticent will find themselves under great pressure to "share." If the idea that death and dying provide new opportunities for self-improvement becomes common currency, chronic underachievers will find themselves facing one more opportunity for failure. Not "getting off" on death may become as déclassé as sexual unresponsiveness. Then, perhaps, a "dismal death" movement will rise to wipe the smile from the face of Death and restore the "Grim Reaper" to his historic place of honor.[6]

Lofland's book will remain relevant for all these specific cautionary reasons, and many more. I hope that in another four decades *The Craft of Dying* is republished for that historical moment's own Happy Death Movements, especially the ones that still evoke the death taboo enemy in order to evangelize a getting-off-on-death gospel. The irony, of course, is that Lyn Lofland showed us all how easy it is to talk about death and dying without ever transgressing any taboos, and she did this forty years ago in a book that the MIT Press had the good sense to republish.

On further reflection it becomes clear that most Happy Death Movements just can't help themselves when it comes to constantly talking about this taboo that isn't actually true. Why? It makes them feel useful. Lyn Lofland would likely say that's okay.

In the face of dying, Death doesn't really care.

7/29/2018

Watching My Sister Die—Fly Faster
I realized little sister that I first wrote about your death on
April 29, 2018
    so few months before today
July 29, 2018
In a year when we spent our final birthdays together
When I turned 45 and thought that maybe you'd live longer
But you didn't
And I'm not surprised
Because I know how death works.
It's my superpower, as you once noted.
So I'm on a second plane now, little sister.
To Italy, where you died
where I realized you'd die in April
During your birthday.
And one of the hardest things I told you   while holding your
     hand
was that you'd never see home again.
That you would die   where you lay
Increasingly unable to see or hear and robbed of any walking.
It wasn't fair and   these memories certainly made your
death
a little easier to articulate.
Because you never wanted to live that way
calling yourself a troll in the basement
And I totally understand why.
I would never want that either.
But you persisted until you couldn't.
Your heart, so much bigger than any of us deserved,
finally gave out.
Finding me now in this second airplane
staring out the window   willing us off the ground so that I
can finally hold your dead hand
and hug your dead body

and kiss your dead cheek →   giving you all the love
a Big Brother can hold for his little sister
wishing I'd made this trip 48-hours sooner but knowing
you heard me say goodbye three-weeks ago, knowing as I told Santa
        Maria
that this might be the last time we see you alive
And it was.   The last time.
I keep pausing, little sister. staring. wiping away tears. searching.
looking. Wondering what happens next because you not being here
was never part of the plan.
I'm on my own now. The   last Troyer. And how dare you do
        this.
Stopping staring again.   Wandering through these
        words   to
try and understand why this plane isn't up in the air already
bringing me to your dead body. Because if I had one
        wish,   right now, it would be you next to
me on this airplane and both of us yelling   as loudly as
        possible.
FLY FASTER! FLY FASTER! FLY FASTER!

3   The HIV/AIDS Corpse

No group of professionals are more aware than funeral directors that Americans are dying prematurely of HIV and AIDS related illnesses. Not only have they had to learn how to help survivors cope with their grief, but they have had to deal first-hand with the inherent risks in dealing with possible contagious remains.

—from *America Living with AIDS* by the National Commission on Acquired Immune Deficiency Syndrome, *The Director*, January 1992

A New Postmortem Situation: The HIV/AIDS Corpse

In the twentieth century's waning years, the supposedly stabilized human corpse took a sudden turn, and in June 1985, the National Funeral Directors Association[1] (NFDA) circulated a memorandum to its affiliated members that detailed "AIDS Precautions for Funeral Service Personnel and Others." The single page document, marked "IMPORTANT" across the top of the page, attempted to answer common questions from members of the funeral industry regarding dead bodies produced by HIV/ AIDS. The opening paragraph of the memorandum provided the following statement: "There has been much publicity given

AIDS (Acquired Immune Deficiency Syndrome). This disease has led to a number of deaths with more expected. According to numerous reports, those most likely to be infected are homosexuals and intravenous drug users."[2]

Since the funeral industry is one founded upon the handling of corpses, the lack of understanding about HIV/AIDS in America at that time meant that the industry's fiscal reliance on dead bodies had suddenly become significantly more complex. The memorandum went on to state, "The growing publicity these cases are getting have some funeral directors asking for advice not only as to embalmers, but also others involved in the care and transportation of an infected body prior to and after embalming. And, what if the survivors want a funeral with a body present but do not want the body embalmed?"[3]

In response to this question, the NFDA suggested eight "embalming precautions" developed by the director of pathology at New York City's Memorial Sloan Kettering Cancer Center and the Queens Medical Examiner's Office.[4] The eight precautionary steps listed what kinds of protective gear the embalmer should wear (goggles, shoe covers, double rubber gloves, etc.) as well as the kind of bleach solution required to clean dirtied instruments. The most historically telling point in the memorandum is the eighth and final precaution on the list: "If the body is being viewed, the family should avoid having physical contact with it." That families should avoid touching the HIV/AIDS corpse is indicative of the broader biomedical confusion and uncertainty caused by AIDS during the 1980s.[5] While the memorandum does not explicitly state *why* funeral directors should prevent families from touching HIV/AIDS bodies, the implicit suggestion was clear—the body killed by HIV/AIDS was, at that time, a biologically dangerous corpse.[6] What the memorandum also makes clear is that HIV/AIDS produced a new kind of dead body during the early 1980s—a human corpse created through a largely misunderstood disease, and, according to the logic of the memo, by the socially abject activities of intravenous drug users and homosexuals.

Yet, the HIV/AIDS corpse was more than just a dead body with socially deviant, abject qualities. This new kind of corpse became both a product of American cultural politics and a direct challenge to the nineteenth-century technologies created to standardize the modern human corpse. As the NFDA memorandum suggests, this new kind of corpse was challenging the transformative powers of mechanical embalming that purported to technologically stabilize all dead bodies. Since the middle of the nineteenth century, the power of mechanical embalming was that it stopped the dead body from rapid decay and transformed the corpse into a more stable, publicly viewable kind of body; the HIV/AIDS corpse undermined the biomedical power of that practice. In a very short series of steps, the HIV/AIDS corpse quickly became the epicenter of a supposedly socially deviant disease, and it marked the sudden impossibility of postmortem rehabilitation through proper embalming by the American funeral industry.

Examining the HIV/AIDS corpse's social and political productivity within the American funeral industry during the 1980s and early 1990s is revealing. This is not a polemic against the American funeral industry; rather it is important to examine how this particular group of professionals made sense of HIV/AIDS through some serious industry-wide challenges. Moreover, the larger institutional changes experienced by the American funeral service industry, resulting from the emergence of the HIV/AIDS corpse, were substantial. A former University of

Minnesota Program of Mortuary Science[7] embalming instructor, Jody LaCourt, reflected on that historical moment: "In the early nineties, which is when I began my funeral profession, I was witness to funeral directors who 'freaked out' knowing they had to embalm a body that was HIV positive or died from AIDS, and some funeral directors would go so far as to refuse to embalm that body. They did not want anything to do with an HIV/AIDS body. These attitudes were extremely disconcerting."[8]

The hard-line refusal of some funeral directors to embalm HIV/AIDS corpses became such a concern in the mid-1980s that the National Funeral Directors Association Board of Governors made the following policy statement on June 23, 1985:

> We know that some funeral directors and certain funeral service establishments in the U.S. are refusing AIDS cases. Some firms allegedly will not embalm the bodies of AIDS victims. In some cases, firms which will serve the families of AIDS families will do so only if the body is cremated immediately and only if no viewing and/or visitation are permitted. All of this is true even though these actions may directly contradict the profession's previously stated views on the importance of embalming. So strong is the fear of AIDS that many in funeral service apparently are not concerned that this "trend of refusal" may impair the professional image of the American funeral director. ... If this trend of refusal continues, especially with the help of those in funeral service, we could be encouraging higher rates of direct disposition[9] [of the corpse]. After all, what about deaths due to other highly contagious diseases? Don't they also pose a risk? If we encourage direct disposition for AIDS victims, shouldn't we encourage it for others as well? If direct disposition is aided and abetted by funeral directors, there will be little incentive for the public to accept and encourage those in funeral service who support the value of funeralization.[10]

One of the funeral industry's central HIV/AIDS information resources during the 1980s and 1990s was *The Director*—the monthly publication of the National Funeral Directors Association. *The Director* has been regularly published since 1957 and remains in print today. Throughout the 1980s and into the 1990s, *The Director* published a series of articles and commentaries on HIV/AIDS and its implications for the American funeral industry. The September 1987 issue of *The Director* was a particularly important publication on this score; it was the first time that an entire issue wholly focused on HIV/AIDS in American funeral practice. The issue's cover (see figure 3.1) was a visual testimonial to how the epidemic functioned at that time within the industry. Test tubes containing various kinds of fluids float across the cover, and an asymmetrical Erlenmeyer flask holds a multipronged orb (which bears an uncanny resemblance to a World War II Naval mine). The orb itself encapsulates a container of DNA, so the double helix suggests, and in the event the reader is unsure what the multipronged orb might be, the caption "AIDS virus" appears inside the bottom of the flask. American funeral directors were engaged in "Total War," to borrow a key HIV/AIDS metaphor from Catherine Waldby.[11] The articles within the September 1987 issue, listed by category underneath the cover image ("AIDS Experts," "Serving All Families," "AIDS Liability," etc.), also convey the scope of funeral directors' concerns at that time. Lastly, the cover's front and center positioning of the word "AIDS" leaves no doubt about either the urgency or severity of the situation confronted by the funeral industry.

The September 1987 issue begins with a new kind of old problem for American funeral directors. When dead bodies produced by complications from HIV/AIDS began showing up in funeral homes, the "threat" of viral contamination was, in theory, another part of the job. Infectious diseases of all kinds

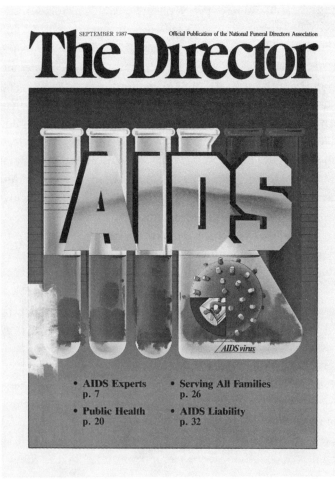

Figure 3.1
The Director September 1987 cover. *Source:* Courtesy of the National Funeral Directors Association.

routinely produced hazardous dead bodies, and those corpses forced the funeral industry to adjust its practices. Professional embalmer and embalming science historian Robert Mayer argues in *Embalming: History, Theory and Practice* as follows: "Deadly flu, polio, AIDS, tuberculosis, and drug-resistant strains of virus, and bacteria have all challenged the embalmer to practice sanitary techniques to protect personal and community health. In spite of these diseases the embalmer continues to practice a skill that allows one last comforting look on the face of loved ones at their ceremony of farewell."[12] Yet for all the other historical examples Mayer cites wherein infectious disease produced dead bodies that required special attention, HIV/AIDS proved to be a very different kind of epidemiological threat. Not only was HIV/AIDS historically different for the funeral industry in terms of the populations it initially infected, so too was the industry's confusion surrounding what to do with the seemingly contagious human remains.

In June 1986, one year before the special issue on HIV/AIDS, the NFDA ran a nine-page article in *The Director*, entitled "AIDS—All the Questions All the Answers." The article's final question is a telling example of the confusion HIV/AIDS was creating for both funeral homes and the people trying to arrange funerals for loved ones:

> **[Question] 86.** *Can funeral directors in New York State refuse to embalm victims of AIDS?*
>
> **Response:** There is no New York State law that can require a funeral director to accept an AIDS victim. Embalming also is not required by law. The state has provided AIDS safety guidelines to funeral directors and embalmers. While some funeral establishments have refused to accept AIDS victims, there are sufficient firms available who will do so.[13]

And yet, as the NFDA Board of Governors asserted, the institutional history and future livelihood of American funeral directing required equal access to postmortem services for all dead bodies. As Clarence Strub and L. G. Frederick argue in *The Principles and Practice of Embalming* (1989): "In time of epidemic or catastrophe, it is our professional obligation to serve and to protect; no matter what our professional jeopardy may be. An embalmer cannot refuse his professional services to a family and ultimately society, just because the deceased died of some communicable disease, such as A.I.D.S."[14]

HIV/AIDS was, however, different, and the bodies produced by it were both the Same and terrifyingly exotic. Paul Rabinow, in *Essays on the Anthropology of Reason*, builds upon Michel Foucault's *The Order of Things* by suggesting that "we need to anthropologize the West: show how exotic its constitution of reality has been; emphasize those domains most taken for granted as universal."[15] And the emergence of the HIV/AIDS corpse did just that; it destabilized notions of universal death, i.e., that in death American dead bodies were mostly the same. The American HIV/AIDS corpse was actually made doubly problematic by its exotic "Otherness" which was explicitly local and domestic. While American funeral practices had been historically quasi-homogenous (different religious rituals being the key exceptions), the emergence of HIV/AIDS produced whole new sets of postmortem rules and regulations that institutionalized extreme forms of homogeneity. The universalization only occurred, however, after the heterogeneous HIV/AIDS corpse disrupted most concepts of a normal, dead body.

Technologies of the HIV/AIDS Corpse

In an article written for the September 1987 issue of *The Director*, Robert Mayer summarized the effects of HIV/AIDS on the American funeral industry. His article, titled "Offering a Traditional Funeral to All Families," is a compendium of public health updates, new health code regulations, and suggested changes to embalming procedures. Mayer opens the article by commenting on the pervasive sweep of the virus across the country: "In looking at AIDS from the standpoint of funeral service, we begin to realize that the disease has spread far enough through our population that funeral homes throughout the entire United States are being asked to provide service for families who have had a relative die from this deadly killer."[16] The article proceeds to present the responses of several unnamed funeral directors when asked how they would handle the preparation of the body if presented with an HIV/AIDS corpse: "One funeral director stated he would have to charge $15,000 to handle an AIDS funeral because the entire preparation room would have to be refurbished. Another funeral director said he considered an AIDS victim to be the same as a highly radioactive body. ... In fact, he stated he would not even accept a call from a family whose relative had died of AIDS."[17] And while there was certainly a shared anti-AIDS corpse sentiment circulating within the industry, it is important to note that other industry members agreed to embalm HIV/AIDS corpses without such overt hesitancies.[18]

That said, the unnamed funeral director's analogy in Mayer's article between the "AIDS victim" and the "highly radioactive body" deserves more development. The underlying premise of the analogy points to specific precautionary measures taken by the funeral industry when embalming a body that has been exposed to any kind of radioactive substance. What makes the analogy significant is how much it reveals about the politics of handling an HIV/AIDS corpse in the 1980s. In the post-WWII decades leading up to the emergence of HIV/AIDS (and in large part because of Cold War politics), a corpse exposed to radiation had been regarded as one of the most dangerous kinds of dead bodies for the funeral industry. The radiation exposure could come from cancer treatments, but it might also arrive in much larger doses from industrial or workplace accidents.[19] Most importantly, these different exposure scenarios all meant that even after an individual had died (from whatever cause), his or her dead body could in fact still contain radioactive isotopes used in medicine or could emit unhealthy levels of post-accident radiation.

The emergence of the HIV/AIDS corpse momentarily displaced the radioactive corpse as the major health risk confronting funeral directors. Indeed, the analogy succinctly illustrates the confusion and the stigma surrounding AIDS that turned these bodies into misperceived biohazard risks. The funeral director's remark is additionally significant since it suggests that the extreme precautions taken to handle radioactive bodies should also be taken when handling the HIV/AIDS corpse. One of the radiation exposure precautions stipulates, for example, that a funeral director should not handle the corpse (even while wearing protective lead-lined gear) until "a radiation safety officer has certified the body as safe because of the possibility of gamma radiation."[20] The analogy's central point is that the HIV/AIDS corpse was equivalent to the most dangerous and most labor-intensive public health hazard a funeral home could accept at that time.

In retrospect, the potential biohazards represented by radiation victims and HIV/AIDS corpses were scientifically incomparable. A highly radioactive dead body poses a far more serious threat to public health than the HIV/AIDS corpse, yet such scientific differences were not readily discernible within the industry at the time.[21] Most dramatically, the funeral director's analogy suggests that merely being in the same room with the HIV/AIDS corpse meant a funeral director would be exposed to potentially life threatening conditions. It also illustrates how central the combined technologies that classify, organize, and physically alter the human corpse are when confronted by an epidemic such as HIV/AIDS.

The term *technologies of the corpse* is partially derived from Michel Foucault's articulation and theorization of what he calls "human technologies." Foucault describes these human technologies as "the different ways in our culture that humans develop knowledge about themselves: economics, biology, psychiatry, medicine, and penology."[22] These technologies are used, Foucault suggests, to produce knowledge and understanding about the self in relation to others. This knowledge production, he argues, can be understood through these four interconnected categories: "(1) Technologies of production; ... (2) technologies of sign systems; ... (3) technologies of power; ... (4) technologies of the self."[23] Of course, it is not only the living body that is subject to these technologies. Through its control over the human corpse, for example, the modern American funeral industry can also be understood as deploying each of these technologies within a postmortem arena. But these technologies are more than just machines; they are ways of thinking about the productive possibility of the corpse and the avenues opened up by these technologies for changing the entire organic structure of the dead body. These technologies of the corpse also then encompass the machines, politics, laws, and institutions controlling the dead body.

Although all four technologies outlined by Foucault are relevant to discussing the productive potential of the HIV/AIDS corpse, the focus here is on his *technologies of the self*.[24] This theoretical use of the technologies of the corpse, which is a step away from Foucault's exact wording on human technologies, frames already existing postmortem tools in a way that is otherwise helpful and useful for living individuals to transform the corpse's "postmortem self." And while it is true that the dead body appears absent of any "self" or "subjectivity," the living individuals that surround a human corpse produce countless (often competing) narratives about a deceased person. The self does not die with the person, rather it transforms into a new form of externally controlled, postmortem subjectivity. These changes imposed upon the dead body are thus a subtler, and by extension, more controlling form of technology, since a dead person lacks the physical ability to dispute most enacted transformations.

HIV/AIDS corpses presented two key problems when transforming the dead body. The first was the direct challenge posed by the virus to the institutional use of embalming machines and to the technicians operating those machines. The second problem was the challenge to the external control of an individual's postmortem identity. Many of the institutions that physically altered the dead body for public viewing were simply refusing the work. If the corpse was left unembalmed and aesthetically unaltered, then obscuring the presence of HIV/AIDS in the dead person was all the more impossible. The invisible nineteenth-century technologies that produced new postmortem conditions through embalming were not only being made visible but also being rendered ineffective when dealing with HIV/AIDS. The HIV/AIDS corpse momentarily called into crisis the theory and practice of human control over how death visually appeared.

Universal Corpse Technologies

During the 1980s and early 1990s, the American funeral industry developed and received new rules, guidelines, and technologies from the federal government for the postmortem handling of infectious bodies, such as HIV/AIDS corpses. Those embalming precautions remain intact today and have not changed in any drastic way since the early 1990s.[25] One organizational system that the funeral service industry uses requires "tagging" a corpse when a person dies.[26] This tagging system, put in place prior to the 1980s, is used to readily identify the corpse's cause of death. The act of attaching the tag to the dead body, either in the hospital or at the morgue, is the first step in categorizing the corpse for postmortem handling. In the September 1987 issue of *The Director*, Mayer used the following anecdote to explain the institutional importance of tagging in response to infectious diseases: "I prepared the body of a young female a few months ago. This young girl had a horrible weeping lesion in the right eye orbit. Two hours after the body had been brought to the funeral home, the hospital called to inform us the girl had a bad case of herpes. The bottom line is simply that today we must prepare every body practicing maximum sanitary measures."[27] Unless a postmortem tag is in place before the corpse arrives, the embalmer must assume that an infectious disease produced the body.

In 1991, in response to precisely these kinds of health concerns, the US Department of Labor and the Occupational Safety and Health Administration (OSHA) created uniform federal rules for the handling of all dead bodies. The rules were known as "universal precautions."[28] It was during this transitional moment that a new kind of postmortem universalization was indeed being created for the dead body. Mayer explains that using universal precautions means an embalmer will "treat all human remains as if they were infected with HIV, HBV [hepatis-B] or other pathogens. In other words, the embalmer should treat all bodies with the same caution that would be applied for extremely hazardous, potentially fatal infections."[29] The following excerpt from a long essay by South Dakota funeral director Dalton Sanders will give the reader an idea of the labor involved in treating all human corpses "as if infected with HIV, HBV, or other pathogens" in order to embalm the body. The article, "Err on the Side of Caution," appeared in the April 1997 issue of *The Director* and demonstrates how the confusion surrounding HIV/AIDS cases in 1987 was transformed, ten years later, into a series of rigorous, meticulous procedures. Sanders's account describes the handling of a corpse when following the OSHA universal precautions:

> My assistant and I wore nonpermeable gowns, heavy autopsy gloves, masks, safety glasses and shoe covers. Upon our return to the funeral home, we placed the body on a table after it was sprayed with a disinfectant. We then cleaned the cot with a bleach solution, and let stand to air dry, and then sprayed it again with a disinfectant. We first treated all open areas with a liquid preservative chemical, and then packed the nasal cavities and mouth with cotton soaked in the liquid preservative. Then, the entire body was washed with an antiseptic soap. After the initial body disinfection, we chose to use a very high index fluid mixture to ensure penetration would be effective,

and chose the right common carotid as the point of injection. The embalming procedure itself went very well, with results that were comparable to a standard embalming. We paid particular attention to any splash or spillage, and cleaned it immediately with cavity fluid. All instruments were placed in one area and were disinfected with a bleach solution. We aspirated immediately following the operation to cover internal areas quickly with 32 ounces of [body] cavity fluid. Any areas that were open type lesions were covered with a cauterizing solution, and the body was then placed in full coveralls with the wrist openings tightly taped shut. The body was then sprayed liberally with a topical spray and covered with a sheet. Prior to dressing the deceased, we dressed as we had for the embalming stressing universal precautions, because the organism [MRSA] is transmitted by direct contact. Cosmetics were applied, (with the understanding that the brushes were soaked in a disinfectant upon the conclusion, and the handles were bleached), and a veil was used to discourage direct contact with the deceased.[30]

What is significant about this description is that the corpse was not in fact infected by HIV/AIDS, but given the ambiguity of the cause of death it had to be treated as though it was a hazardous case. By universalizing the potential hazard, the American funeral industry adapted to the complications produced by the HIV/AIDS virus with a system of institutional controls that made every human corpse a potential biohazard. In effect, and through the institutionalization of universal precautions, concerns over the HIV/AIDS corpse established a new standard for all bodies through those very same practices. Simultaneously, these processes came to treat *all* human corpses as though they were potentially epidemiologically hazardous and abnormal, even if a postmortem exam said otherwise. The result of these arduous procedures is that all bodies suddenly become universally safe to handle, touch, and view after cosmetics are applied. These institutional protocols are also an example of Foucault's point that technologies for controlling the abnormal individual are ones that give rise to "theoretical constructions with harshly real effects."[31]

The combined technologies of the corpse subsequently conducted the HIV/AIDS corpse through a two-step, oftentimes severe process. First, the technologies were used to categorize the HIV/AIDS corpse as an abnormal body through an array of regulations and guidelines dictating how the body should be handled. After the technologies were used to define the abject condition of the HIV/AIDS corpse, the same technologies were then applied to bring the dead body back into normal conditions. The postmortem individual connected to the HIV/AIDS corpse consequently became an aberration through the deployment of technologies that were then redeployed to make the dead person human again. *The HIV/AIDS corpse was a monster made, not a monster born.*[32]

And while the HIV/AIDS corpse presented problems for many funeral directors during this period, it also represented an opportunity to reorder the technologies of the corpse in new ways. One of the ways that some professional embalmers overcame the perceived health threat associated with the HIV/AIDS corpse was to effectively heroicize the work being done when embalming the HIV/AIDS corpse. In a July 1985 article in *The Director* entitled "AIDS—Identification and Preparation," Jerome F. Frederick, then the director of chemical research for The Dodge Chemical Company (a major American embalming fluid manufacturer), argued: "Perhaps no threat has ever appeared on the health horizon like the one facing the professional embalmer today. The disease threat is AIDS ... and, despite constant research since it was first discovered among men in the homosexual community, it continues, after five years, to baffle our sophisticated medical establishment."[33] What both Frederick and Mayer suggested in those years was that even though the technologies made available to the professional embalmer and funeral director appeared threatened by HIV/AIDS, the tools must always overcome any challenge to altering the dead body. In order to preserve the theory, practice, and economic livelihood of those technologies, the transformation of the dead body needed to occur in spite of the perceived health risks and aberrant qualities projected onto HIV/AIDS corpses. In other words, the technologies of the corpse required reassertion even though funeral directors or embalmers were personally ill at ease around the dead bodies of, say, homosexuals or intravenous drug users. The entire point of deploying the technologies of the corpse was to enable living bodies to control dead bodies, not vice versa. Technologies preserving the corpse could therefore not be rendered useless by the very thing that the technologies were originally developed to manipulate: the dead body.

There is a final point worth noting on the transformative power of these technologies. Even though the HIV/AIDS corpse was a socially and medically different dead body than previously encountered by the technologies of the corpse, that body still possessed a transformable "self" if it was given proper care. So, when the HIV/AIDS corpse first appeared, embalmers and/ or funeral directors were no longer strictly nineteenth-century preservation technicians. The technologies' institutional administrator had gained the power (often through the direct orders of the family) to transform the corpse into an entirely different kind of person. Even though the HIV/AIDS corpse posed a challenge, an extreme health risk in the minds of some funeral directors, the postmortem transformation of the person who died became all the more possible with the technologies' reordering.

After dying, the transformation of the dead body meant entire histories of the deceased person's life could be obscured, even left unknown, given the power of the technology to alter an individual's attributes. Who and what that "postmortem self" became then meant embalmers and funeral directors using the technologies of the corpse had great power to transform the HIV/AIDS corpse back into a human being much different in death than perhaps ever in life.

The emergence of the HIV/AIDS corpse in the 1980s turned the dead body into something quite different than it had been a century before for American funeral directors. Compounding these problems were funeral directors who refused to handle these bodies, meaning the technologies invented a century earlier to standardize the human corpse were suddenly useless without an operator. The institutional history of postmortem preservation, stressed by the American funeral industry, also seemed at odds with the fear many embalmers expressed at the very idea of touching bodies produced by HIV/AIDS. The previous century's postmortem technologies and ways of thinking about dead bodies did not seem to initially work when applied to HIV/AIDS corpses. This epistemic shift redefined the dead body, so much so that different postmortem technologies at the levels of both thought and practice were developed and deployed by the American funeral industry in order to control the HIV/AIDS corpse.

The HIV/AIDS Corpse and Queer Politics

In *AIDS and the Body Politic*, Catherine Waldby outlines and critiques the scientific rhetoric defining the AIDS epidemic, as well as the characterizations of human sexuality attached to the virus.

One of Waldby's central arguments defines "AIDS as a symptom, not of the activity of a virus, but of a particular moment in the history of sexual politics."[34] The American funeral service industry found itself enmeshed in that historical moment, and the projection of human sexuality (rational or not) onto dead bodies forced an institution-wide and often very heteronormative redefinition of the human corpse. This redefinition resulted in the HIV/AIDS corpse becoming a politically productive body that altered both institutional codes and technological conventions.

In the conclusion to *AIDS and the Body Politic*, Waldby cites Judith Butler's suggestion that terms such as "queer" and "queer politics" raise "the question of 'identity,' but no longer as a pre-established position or uniform entity; rather, as part of a dynamic map of power in which identities are constituted and/or erased, deployed and/or paralyzed."[35] Waldby expands Butler's argument with the suggestion that "perhaps queer is also an attempt to rework relationships between identity, contagion, and death, an ethical task of the self which the advent of AIDS has made so urgent."[36] Waldby's ethical task also suggests that the HIV/AIDS corpse needs to be recognized for its significant productivity in queer politics, that is, that it radically reworked death and dying's social and political dynamics by challenging the technologies that pathologized a person who died from AIDS. In death, the HIV/AIDS corpse produced a moment of paradoxical "postmortem agency" for an individual who might otherwise have remained silent.

In a 1992 essay for *The Director* on precisely this issue of pathologization, social worker Michael Hearn wrote a testimonial piece about the conflicts between different groups (families, lovers, friends, etc.) when acknowledging the identity, contagion, and death of a person who died of AIDS. Hearn describes attending the HIV/AIDS funeral as a regular occurrence during the 1980s and a situation where the body in the casket was fixed with new and conflicting meanings. He writes: "As a gay man, I've watched the issue of AIDS become a springboard for the hate of an entire population. I've cried the tears of a lifetime at funerals where the dead are remembered for having wasted their lives. Where the casket serves as the dividing line (relatives to the left, "us" to the right). ... And I've listened to the litanies, the soliloquies, and monotonies where illnesses are covered up or described with pretty words."[37] A few years later, in a 1996 article for *The Director*, Iowa funeral director Michael Lensing detailed how best to negotiate potential conflicts between groups of mourners who maintained different kinds of relationships with the deceased. He suggests the following tactics: "We are often trying to balance two families at the time of death—the biological and the chosen—trying to satisfy the emotional and psychological needs of both. It may be necessary to divide cremated remains in half to disperse in different locales by different survivors. ... Be sure to find out who is legally in charge."[38]

The legal distinction between the chosen and the biological family often meant that the lawful standing of biology outweighed kinship choice. This also meant that the law could, and often would, exclude a chosen family from placing any claim to the biological family's dead body. The legal responsibility of the funeral service industry to the biological family, as mandated by the state, potentially risked the chosen family's complete erasure. The state's judgment regarding who received the bodily remains became less a question of family than a question of property ownership.[39] While the biological family may have had legal rights to the body, those who cared for the deceased were often left with no legal standing.[40] Kath Weston underlines the fictitious, but no less real distinctions between a chosen and a biological family structure when she argues that "the standard-ized 'American family' is a mythological creature, but also ... an ideologically potent category."[41] These family group reconcep-tualizations marked an important moment for the HIV/AIDS corpse and queer politics, as the legal right to the body high-lighted questions of sovereign power and technologies of con-trol over the self. Politically active queer politics, endowed with such urgency by HIV/AIDS during the 1980s, also illuminated the broader conflict between an *ethical claim* and the *legal right* to possession of postmortem human remains.[42]

In the *History of Sexuality, Volume I,* Foucault begins the final essay, "Right of Death and Power over Life," by explaining "for a long time, one of the characteristic privileges of sovereign power was the right to decide life and death."[43] The pathological positioning of the HIV/AIDS corpse offered a new opportunity for sovereign authorities, that is, the people with a *legal right* to the corpse, to exercise power over the individual well beyond the individual's death. The power exercised was not so much the sovereign's decision about life and death since the bodies involved were already dead. But even if the question of death was answered, it did not mean that the dead person involved was without the possibility of living another kind of life. That other kind of life, one unfettered by the dead self associated with HIV/AIDS, is best demonstrated by the technologies of the corpse transforming the human deviant into a new kind of nor-malized individual.

If queer politics is to forge a different kind of identity and make sense of living, per Waldby's suggestion, then these formu-lations about life must also make sense of death. The HIV/AIDS corpse is that *thing*—neither a complete subject nor entirely an object—that refuses to go silently, no matter the controls exercised by sovereign power under the legal right of the state. The HIV/AIDS corpse might one day be recognized as having had a paradoxical agency that embodied Waldby's ethics and challenged how state power exercised control over dead bodies.[44] The irony of this situation is that the socially abject, HIV/AIDS corpse will have caused these changes for the postmortem politics of all American bodies.

Postscript on the Temporality of HIV/AIDS

In 1992, the Infectious/Contagious Disease Committee of the Funeral Directors Services Association of Greater Chicago commissioned a study to determine how long the HIV virus remained active in a body after the host died. The results of the study, published in the January 1993 issue of *The Director*, are well worth mentioning here, if for any reason, the study found that "for unknown reasons the viability of the virus appears to be time-dependent following death," that in twenty-one of forty-one subjects, scientists "were able to isolate the virus up to 21.15 hours after the patient's death. No virus was found after that time, and refrigeration of the deceased did not significantly affect the virus's survival rate."[45] Assuming that a new mutation of the virus has not invalidated this study, an HIV virus that expires after less than twenty-four hours changes (biomedically speaking) the status of the HIV/AIDS corpse. If a corpse ceases to be infectious after that amount of time, then that body should, in theory, remain nonthreatening for the funeral. On a sociocultural level, of course, that reality has not and likely will not register in the near future for the HIV/AIDS corpse. Once the technologies of the corpse mark a body with HIV/AIDS, undoing those signifiers is difficult at best.

Yet a point left to ponder is the temporality of the HIV/AIDS epidemic.[46] Assuming that the epidemic will continue for some time, then the institutional changes that altered both the funeral industry and how funeral directors handle human corpses will most certainly endure. In 2020, the same universal precautions created during the early 1990s remain standardized procedures and underscore the central significance of the HIV/AIDS corpse: the invention of an entirely new kind of dead body that brought the technologies transforming these bodies into public view and debates.[47] Of course, the institutional arguments taking place within the funeral industry were not always in public view, but the resulting changes most certainly appeared on view for the public.

The HIV/AIDS corpse was, in so many words, a useful body for the technologies of the corpse, presenting entirely new opportunities for changing how previously used machines interacted with the dead body, as well as defining the limits of sovereign authorities' power. More than anything, the treatment of the HIV/AIDS corpse demonstrates how in flux the concept of the human corpse can become when seemingly inert dead bodies begin destabilizing the previously held regimes of knowledge defining death.

In closing, it is important to point out that through popular imagination, a general lack of knowledge, and certain funeral industry practices, the HIV/AIDS corpse became the most threatening dead body of its time. The productive potential of the AIDS epidemic can, as a result, be partially located in the invention of this new kind of corpse. In hindsight, it is clear that only an epidemic as pervasive, resistant, pathologized, and deadly as

HIV/AIDS could have initiated the funeral industry's massive institutional changes to well-established technological practices. What the temporality of HIV/AIDS may yet produce is something beyond a corpse—a dead body that maintains a politically active position without the privilege of vitality. The technologies of the corpse are always at work classifying, organizing, and physically altering the corporeality of the dead body. Changes produced by those technologies, including the codification and normalization of abject corpses, have undoubtedly drawn attention to the fallacy of control often attached by the public to modern death practices. The early decades of HIV/AIDS in America may become recorded as an historical moment wherein corpses emerged within the politics of death as a new kind of dead body—one that confounded the technologies of regulation and control in everyday life.

8/06/2018

Watching My Sister Die—#19. Julie's Funeral
I can't cross your funeral off the list, little sister.
And I won't.
I knew this even before I wrote it down
on my weekly list
Like an ordinary task   right after 18. Gym.
It's taken me these past six-days to actually
    open the pen cap and stare at this page
Stare out this airplane window
leaving all of you behind
Having touched your dead hand and kissed you goodbye
when I arrived.
Finding ourselves, like we did as children, in a mortuary
and feeling completely at home holding your   marble white
    skin
Your face frozen now with release.
And I brought Mom & Dad to you → keeping my promise
Making sure that they were ok.
Because little sister I feel your absence now in
the silence
In the quiet of you not yelling at me about my life.
So I laugh little sister and I cry and I look at photos that
represent us in multiple ways. The decades of our lives.
The smiles at the end that hid the pain.
The promises you made me make to look after your children.
And Mom & Dad, who I know will join you sooner than I want.
Reminders of you throughout your house in all the little things,
the clothes, the box of tampax,   the coffee machine
that you used right up until you died.
The bed where you laid, in the basement, when you could no
longer walk. Holding my hand as we   talked about dying.
Me sleeping in this bed, when I arrived, after you died
smelling your hair, the same hair I smelled at the mortuary,
on the pillow.

I watched you die little sister. Over 365 Days.

Realizing now that I could have done more to expedite your
pain relief.

Intervening sooner. At your birthday party when I knew that
there would be no more.

But I didn't and I'll forever wonder why

what part of my personal life overrode my professional knowledge.

Then being there for   your husband when he needed
        me.

Going with him to file your death paperwork

All the forms that needed signing to say that you no longer lived.

    staring at those forms with your husband and watching
him sign them all.

On his birthday and on that day for the rest of his life.

Here I go little sister. Up into the air. Leaving where you've already
        left.

Everything's different now, little sister, and strangely the same.

    A vast blurring of   how I understood death.

And in this Patient Zero moment I realize now that I knew very
        little until   I

stared at #19. Julie's Funeral.

4 Plastinating Taxonomies

> In every major city in the world there are countless museums that exhibit the products of human culture, sometimes featuring highly unusual themes. However, there is not a single museum about humans themselves—an institution that exhibits the anatomy of healthy and unhealthy human bodies in an aesthetically pleasing way using authentic specimens.[1]
>
> —Gunther von Hagens, *Donating Your Body for Plastination*

The Human Museum

In an interview with National Public Radio's *All Things Considered* entitled "Cadaver Exhibits Are Part Science, Part Sideshow," *Body Worlds* creator Gunther von Hagens discussed his new millennium plans for a future dead body exhibition. The program's reporter explained that in 2006 von Hagens sent questionnaires to approximately 6,500 people who wanted to donate their bodies to *Body Worlds* after they died. Von Hagens asked some "provocative questions," according to the radio story: "For example, would they consent to their body parts being mixed with an animal's, to create a mythological creature? Would they agree to be 'transformed into an act of love with a woman or a man?' Von Hagens says that on the sex question, the majority of men liked the idea, while the women did not."[2]

Von Hagens's questionnaire should hardly come as a surprise. Since opening in the mid-1990s, the *Body Worlds* exhibitions have generated equally large ticket sales and audience numbers—forty-four million visitors and counting as of 2019.[3] What von Hagens has done over the years is produce a portfolio of work that consistently poses dead human bodies in technologically novel ways, even though many of his exhibitions suggest a connection (however tenuous) with centuries-old anatomical displays. His *Body Worlds* exhibitions succeed by explicitly using anatomical science's history and language to produce popular culture narratives about the dead body. His methods always involve plastination, a kind of dead body embalming technology that he defines as "an aesthetically sensitive method of preserving meticulously dissected anatomical specimens and even entire bodies as permanent, life-like materials for anatomical instruction."[4] Even for the always-industrious von Hagens, however, the exhibition of dead human bodies mixed with animal parts and dead bodies posed in sexual positions sounded far-fetched.

That is, until May 7, 2009, when Gunther von Hagens opened a new *Body Worlds* exhibition in Berlin, Germany, called *The Cycle of Life* (*Der Zyklus des Lebens*). In one section of the exhibition he and his team posed two different pairs of bodies in sexual positions. Von Hagens released a statement in which he explained that the exhibit "offers a deep understanding of the human body, the biology of reproduction, and the nature of sexuality."[5] He also made it clear on the *Body Worlds* website that he wanted to bring the copulating corpses to other cities, such as London.[6]

Each couple consisted of a man and a woman engaged in heterosexual sex. Von Hagens posed the first couple in a sitting position and then sliced their bodies into a thin cross-section that showed the male penetrating the female. The other display involved two, fully formed bodies in which the female corpse was sitting astride the male's body and the female's back was to the male's face. Both couples were in a separate room from the rest of the exhibition, and only viewers ages sixteen and up could enter. Unfortunately, the May 2009 exhibition neither presented human and animal body parts fused together to create mythological creatures, nor has Gunther von Hagens apparently attempted to create these fantastical bodies.[7]

During *Body Worlds*'s over twenty years of existence, it has continually turned the dead body into something new. It is this quest for "newness" (and novelty) that von Hagens is arguably embracing with both *The Cycle of Life* exhibition and his "provocative questions." The most productive aspects of those questions have little to do, however, with the controversies that surround von Hagens and *Body Worlds*.[8] Rather, it is more interesting to ask why positioning the dead body in human sex acts or fusing it with animal parts is itself provocative? Without too much exaggeration, the bodies in *Body Worlds* do everything *but* have sex. Posing dead bodies in sexual positions, perverse as it sounds, is one of the few, common human activities not regularly displayed by von Hagens. Posing dead bodies mixed with animal parts is also not entirely different from the current exhibitions. One of von Hagens's more famous plastinates is of a man riding a horse. This exhibit piece is a prime example of the merger between human and animal bodies. Von Hagens has simply proposed to eliminate the demarcation line between rider and horse in order to create a centaur. Displaying dead bodies having sex is certainly more provocative than fusing dead bodies with dead horses, but both these proposals simply encompass the next logical step (and perhaps conclusion) for the *Body Worlds* exhibitions.

The entirety of *Body Worlds* is itself a provocation, a dare, a direct challenge to look at the plastinated dead bodies. It is no small coincidence that in America, von Hagens exhibits his work most often with science and natural history museums.[9] By placing plastinated corpses in science museums, von Hagens indulges popular culture's fascination with the shocking dead body but he also suggests to the viewer that it is perfectly acceptable to look at abject bodies. Add the prospect of dead bodies having sex to the mix, and a science museum's stamp of approval will only guarantee that the so-called provocation is entirely educational. In a most peculiar way, Gunther von Hagens has given many museums new financial hope by tapping into an older form of cultural shock, that is, the long-running morbid fascination with the human corpse. To reject von Hagen's questionnaire out of hand because it seems too perverse, voyeuristic, or gratuitous misses a fascinating series of arguments. Discussing dead bodies posed in sexual positions and the creation of human-animal mythological creatures offers an opportunity to position von Hagens's overall work among debates in cadaveric anatomy, human taxonomy, and death. It is also far more productive to embrace von Hagens's provocative questionnaire and its possibilities for the technologies of the corpse.

The history of anatomical display, such as von Hagens's, is hardly new and reaches back several centuries. The display of human sexual anatomy, as an educational tool, is very much a part of that history. In *A Traffic of Dead Bodies: Anatomy and Embodied Social Identity in Nineteenth-Century America*, Michael Sappol gives this example: "Anatomical discourse provided a vocabulary in which bourgeois women could speak of their embodiedness, including the sexual body, in a refined, dignified manner, without resorting to overly delicate euphemisms, elaborately indirect allusions, or vulgarity."[10] These displays often involved wax models, drawings, or medical textbooks. Based on von Hagens's recounting of his questionnaire's results, the female respondents clearly did not support the sexual displays, but it is unclear what they found objectionable. The male respondents had no apparent problems with displaying postmortem sex acts, although the numbers on heterosexual versus gay sex would be interesting to see. No further discussion is given at all as to whether or not the respondents objected to fusion with dead animals. Von Hagens's survey group is presumably already inclined to donate its bodies for his use, so it seems fairly reasonable that they are not bothered by *Body Worlds*'s display practices. Even if the underlying presentational concept behind *Body Worlds* remains acceptable, it is the use of cadaveric anatomy in overtly sexual positions that many people find especially objectionable.

When *The Cycle of Life* exhibition opened in 2009, for example, a number of German politicians responded with disgust, shock, and moral outrage. Social Democrats MP Fritz Felgentreu stated, "Love and death are obvious topics for art, but I find it quite disgusting to use them in this way."[11] Green Party MP Alice Ströver proclaimed, "This couple is simply over the top, and it shouldn't be shown."[12] Christian Democratic Union MP Kai Wegner presented a somewhat more pragmatic critique: "I am firmly convinced that [Gunther von Hagens] just breaks taboos

again and again in order to make money. ... It is not about medicine or scientific progress. It is marketing and money-making
pure and simple."[13] Michael Braun from the conservative Christian Democratic Union stated that the dead bodies posed in
sexual positions were "revolting. Hagens rides on a wave of
taboo-breaking and the couple plumbs the depths of tasteless-
ness."[14] To put a dead body on display (even with its anatomy
in full view) is one thing. To put a dead body on display with its
anatomy sexually positioned in/around/and near another body
radically alters that visual tableau. It reduces dead bodies, as von
Hagens's German critics assert, to something vulgar, less human;
the corpses become a great deal more animalistic.

Humans are an anomaly, however, in the animal kingdom, and that is the underlying historical dilemma that Von
Hagens's entire *Body Worlds* project confronts. Carl Linnaeus,
the eighteenth-century inventor of modern zoological taxonomy, explicitly fashioned our animal existence around a bold
presumption: We humans know that we are human. Not only
do we recognize our own humanity, we also recognize that the
other animals are *not human*. What we lacked at the time was a
proper scientific name. In the tenth edition of Linnaeus's *Systema Naturae* (1758), he finally gives human primates a full binomial designation: *Homo sapiens*.[15] Yet Linnaeus's designation for
human primates is a methodological paradox. Giorgio Agamben
uses Linnaeus's central argument in the *Systema's* introduction
to explain: "[M]an has no specific identity other than the *ability*
to recognize himself. Yet to define the human not through any
nota characteristica [a physical trait], but rather through his self-
knowledge, means that man is the being which recognizes itself
as such, that *man is the animal that must recognize itself as human
to be human*."[16]

Being human, then, is based not upon any physical character-istic, such as the binomial designation for our extinct hominid cousin *Homo erectus*, but rather upon our superior cognitive skills. Our genus is *Homo*, just like our closest primate relatives, but our species is *sapiens*, that is, the ones who are self-aware. We *Homo sapiens* are hence defined not by our anatomical structures but by our intellect. When any human being dies, however, intel-lectual ability (as located in the brain) eventually stops function-ing and the "person" attached to that body ceases to physically live. The death of the person also produces a corpse, and that dead body is composed of physical, human anatomy that will begin decomposing if left unpreserved. Death remains a persis-tent physical transformation for *Homo sapiens* that, as animals in the animal kingdom, we cannot currently escape. Per Linnaeus's suggestion, we humans do recognize ourselves, both in life and death, but only so far as this recognition avoids the unsightly presence of the dead body's decomposition. *Homo sapiens* are therefore confronted by the following taxonomical, postmortem dilemma: the unaware human corpse biologically decomposes, and through that process of breakdown it becomes an animal body composed entirely of anatomical *nota characteristica*.

Death physically reduces *Homo sapiens* to that one thing we are supposedly taxonomically superior to: anatomy. This is both a physiological and an ontological dilemma. If in life humans are defined by their cognitive skills, then how is it possible that death so quickly reduces the human being to an animal state? Von Hagens takes this physiological and ontological dilemma an entire leap forward by combining it with the possibility of post-mortem human sexuality. In an unintentionally ironic critique of Linnaeus, von Hagens is making a less-than-subtle argument about human-animal taxonomy. He is taking human anatomy and using it in acts that seem to absurdly demonstrate *Homo sapiens'* cognitive superiority: *Even when dead, human beings can still figure out how to have sex.* But von Hagens is also reducing human sexuality to something that is neither procreative nor purely pleasurable. These dead bodies will have sex without ever knowing that they are physically capable of such a hyperstimulated, Bisga Man-esque activity. Postmortem sex is sex without purpose, function, or evolutionary importance. However, it is something for audiences to stare at and think about in the most fantastical kinds of ways. Von Hagens is inviting the viewer to look at and take total control of an utterly impossible situation—but a situation his questionnaires suggested men strongly supported and women wanted no part in. Even then, the female body is still subjected to a postmortem objectification and sexualization that parallels the everyday lived experiences of many women.

Trying to exert total, technological control over the dead body's decomposition has become a labor-intensive project for the modern First World. What that constant labor produces, however, comes with its own limitations. A possibility that leads to this question—How far are people willing to go to prevent this postmortem human-animal slippage? To what didactic extremes can the dead body be taken to appear still alive, laboring, and, most importantly, human? One of the best examples of these didactic borderlands (for both educational and entertainment reasons) is the work done by Gunther von Hagens in the *Body Worlds* exhibitions.[17] But even the ever-imaginative Gunther von Hagens faces a potential limit when making dead bodies look alive, more human than animal, and anatomically active. That limit is human sexuality. Rethinking what constitutes the apparent pathologization of human sexual anatomy is the most productive way to circumvent this same limit.

Posing

> The unlimited shelf life of plastinates makes worthwhile highly detailed dissection work that would have been too time-consuming prior to the invention of the process. After it has been saturated with silicone rubber but before it can be cured, for example, a whole-body plastinate can be first posed as desired. ... Presenting bodily functions in this way sometimes requires positions that reflect certain themes.[18]
>
> —Gunther von Hagens, *Donating Your Body for Plastination*

By plastinating human corpses, Gunther von Hagens asserts an explicit, total technical control over how those dead bodies appear. This is both obvious and entirely obscured within the *Body Worlds* exhibitions. One of the most important actions that von Hagens takes with all of the plastinated bodies is the individual corpse's assigned pose. Without artistic and editorial decisions about how the bodies look (throwing a ball, practicing yoga, playing chess, etc.), no narrative can be created. The preservation and posing of the bodies is done behind closed doors, it takes several weeks, and audiences only ever see the finished products.[19] Von Hagens's posing of dead bodies in sexual acts involves two key concepts: pathology and human labor. Each of these terms deserves further attention.

In order to transform human corpses engaged in sexual activities into didactic displays, the dead bodies must be defined as a return to nature or the normal. This return to nature contrasts with how the plastinated bodies might otherwise be viewed: as a rejection of nature and embodying the ultimate, technological control of the dead body. Von Hagens is very clear in his arguments that the plastination process returns the dead body to nature (albeit counterintuitively) by producing a more natural understanding of human anatomy. The posing of dead bodies, then, involves choices that confront Western attitudes about social decorum and institutional codes of conduct. Von Hagens persistently states that he sees these modern social institutions as repressive: "There is this kind of media influence, and there is a kind of educational influence, which starts in the childhood, 'You should not explore your body,' you know. 'You should not touch your body, masturbation, you should not look on your faeces, er, as a composting machine.' And finally, you are educated so much that you shy away from your body."[20] He also consistently argues that the pathologization of human bodily activity (supposedly abject behavior in particular) occurs well before he poses any cadaveric anatomy. Ironically, von Hagens needs this problematic pathologization to occur before he poses the bodies, since if it has not already taken place, then his *Body Worlds* exhibitions will not challenge various social norms. The societal challenges created by the exhibitions are most assuredly not universal, but they certainly defy enough rules to draw large crowds.

By stripping away its pathological connotations, von Hagens is attempting to create a whole new definition for human anatomy. He is not eliminating pathological possibilities; rather he is embracing them as both necessary and good. This redefinition of the pathological occurs in a retrograde motion: it starts with death and then moves toward life. One of Michel Foucault's key observations in *The Birth of the Clinic* helpfully illustrates how this shift in meaning works.[21] In chapter 8, "Open Up a Few

Corpses," Foucault outlines the development of comparative anatomy in eighteenth-century Europe and England. Foucault's key insight is that during the eighteenth century, pathological anatomy had to be discovered as contributing to death. Odd as it seems today, older forms of clinical care did not understand the cause-and-effect relationship between death and the dead body. The discovery of pathological anatomy created two different kinds of temporality for the human corpse. Foucault explains, "The possibility of opening up corpses immediately, thus reducing to a minimum the latency period between death and autopsy, made it possible for the last stage of pathological time and the first stage of cadaveric time to almost coincide."[22]

What these older forms of pathological time and corpse time lacked, however, was a method to permanently preserve the dead body. Time was a critical factor in making sure that enough information was gleaned from the corpse before it was too thoroughly decomposed. Von Hagens invented his own way around this dilemma, and that technological innovation on nineteenth-century preservation concepts produced a new kind of temporality for the dead body. Through the use of plastination and the posing of dead bodies, von Hagens is creating a whole new kind of corpse time. His preservation process, unlike nineteenth-century embalming, does not simply prepare bodies for funerals or transcontinental shipment; it totally stops biological decomposition. Dead bodies stop functioning like dead bodies.

The transformation of the corpse and its interior into an inert, posed entity reduces any hint of pathological time by prolonging, as von Hagens asserts, the shelf life of the dead body itself. Once the corpse is separated from decomposition via plastination and placed in the posing process, von Hagens controls how pathological categories such as normal and abnormal anatomy visually appear. Pathological anatomy became productive in its originary, eighteenth-century moment because it helped make death visible. Von Hagens needs pathological anatomy to remain productive because it makes his entire career possible.

The conflict between the normal, the norm, and the pathological, as seen in von Hagens's plastinates, is hardly a new topic. In the 1940s, philosopher of medicine Georges Canguilhem explored the boundaries of the norm-normal debate in *The Normal and the Pathological*. Canguilhem explains, "We could say of the two concepts of Norm and Normal that the first is scholastic while the second is cosmic or popular. ... The normal is then at once the extension and the exhibition of the norm. ... A norm draws its meaning, function, and value from the fact of the existence, outside itself, of what does not meet the requirement it serves ... the normal is not a static or peaceful, but a dynamic and polemical concept."[23]

Central here is Canguilhem's observation that the *normal is a simultaneous extension and exhibition of the norm*, as well as his last point—that the normal is a polemical concept. The dead bodies on display in *Body Worlds* are simultaneously normal (as in ordinary, human bodies) and polemical since putting physically active dead bodies on display challenges multiple Western postmortem social codes. These are not dead bodies simply lying in state or even advertising products like the Bisga man did. Von Hagens creates impossibly hyperactive dead bodies. It is precisely this polemical concept, however, that von Hagens uses to transform unaltered dead bodies into exhibition-quality, posed plastinates. So, on the one hand, von Hagens requires the prior pathologization of human sexual acts. This sexual pathologization turns otherwise normal bodily function into something bad, unhealthy and deviant. On the other hand, and even though von Hagens chafes at this social moralizing, the polemical underpinnings of the normal give him vast leeway in establishing a new norm. By using dead bodies, the one kind of human body that audiences recognize as human and that plastination makes completely controllable, von Hagens asserts that human anatomy can defy pathological social conventions through sexual labor. Plastinated dead bodies become counter-intuitively liberated from both decomposition and pathology by demonstrating how normal their sex acts appear. The exhibition of the normal in *Body Worlds* tells audiences that they have gotten the norm all wrong: the dead body is not limited by death, time, or decomposition.

What von Hagens absolutely requires in order to accomplish this conceptual shift is human sexuality. He can display corpses playing basketball all he wants, but if the goal is to really make audiences contemplate human behavior, then sexuality becomes that one kind of quasi-universal, human labor that produces strong responses. But displaying dead bodies having sex also creates a paradox. If human sexuality is in part about procreation (setting aside the theological arguments), then dead bodies having sex represents the total absence of reproductive possibility. It transforms the human sex act, which is endowed with countless kinds of meanings, into a biologically inert, visual tableau. The larger question to ask is this: Which sex acts, poses, kinds of partners, and so on will be displayed? Given that postmortem sexuality between two dead bodies is itself impossible, it seems odd to think through which sexual acts are permissible. That said, the bodies involved are human, so any reference to sexuality comes laden with cultural politics, a situation previously encountered with the HIV/AIDS corpse. The most politically controversial kinds of sexual acts, setting aside illegal behavior such as pedophilia, not only fit within the larger pathological-norm-normal argument but also make the most profound point. If a male corpse and a female corpse can be posed in a sexual position and it is normal, even if it defies the norm for dead bodies, then why cannot same-sex corpses do the same? Perhaps an exhibition displaying bodies with multiple partners or individuals masturbating. How is one kind of dead body labor correct and the other kind wrong when dead bodies do not regularly have sex at all?

This is a kind of human labor that Paolo Virno discusses in reaction to Foucault's concept of biopolitics, or the total management of human life by sovereign authorities. As Virno explains: "[T]o comprehend the rational core of the term 'bio-politics,' we should begin with a different concept, a much more complicated concept from a philosophical standpoint: that of *labor-power* ... what does 'labor-power' mean? It means *potential* to produce. Potential, that is to say, aptitude, capacity, *dynamis*."[24] Capturing this *dynamis* is von Hagens's chief success: the plastinated dead body is not just posed in an active position; it is posed in a dynamic sexual position that through its challenge to social norms and normality becomes active. The case can be made that other poses, such as a basketball player or a chess player, are also active but that those poses do not defy the same conventions. By infusing sexuality into the situation, von Hagens demonstrates that the dead body is pure potential; only a lack of imagination restricts the posing. And it is into the supposedly impossible sexual imaginary that von Hagens takes the dead body. It is sexual labor without reproductive potential, but it is also sexuality without disease, pleasure, or consequence. Dead bodies having sex suggest and present an impossible norm, but a polemically normal impossibility.

Denatured

> In visiting [*Body Worlds*] one perhaps obvious fear can be ruled out: plastinations have been thoroughly denatured just like fossils and even more than mummies, regardless of how fresh and natural they appear.[25] (*Body Worlds* Guidebook)

In many ways, von Hagens's proposed mixing of human and animal anatomy is far more transgressive than posing corpses in sexual acts. This point may strike many readers as counterintuitive, since human sexuality seems a far more taboo topic than creating plastinated centaurs, mermaids, or minotaurs. An exhibition in which audiences see dead bodies posed in sexual positions most certainly grabs a person's attention, but sex is a commodity used for selling many products. What the proposed fusion of human and animal anatomy invokes is a specific human mythology, and it is the mythological's relationship to the taxonomic sciences that truly redefines *Homo sapiens*. The proposed exhibition is not so much an innovation in display technologies as a return to Linnaeus's eighteenth-century discussions on how humans differ from animals. Early taxonomists often struggled to remove mythological creatures from the zoological sciences in order to classify humans.[26] Presenting human mythology in the *Body Worlds* format also entangles the concept of *Homo sapiens* in a certain anatomical fiction inherent to our taxonomic name. We humans most certainly invented the mythological creatures, but we also invented ourselves as nonanimal animals. In fact, all of von Hagens's exhibitions feature *primates,* but the most hairless and pedestrian primates of the animal kingdom.

One primary taxonomic question that emerges from von Hagens's work asks this: What place do plastinated dead bodies have in nature? Von Hagens often argues that the replacement of organic tissue with a nondecomposing exterior (such as a fossil) denatures the dead body. The preservation reduces spectators' "fear" by rendering inert the "fresh and natural" looking corpses. On this point, von Hagens is absolutely correct. A nondecomposing human corpse that audiences know (1) is dead and (2) is not a model does defy the natural order. What von Hagens actually produces through plastination, however, is more than just a denaturing. He is trying to reinvent the entire concept of a natural order, and through this redefinition he is also creating a new taxonomic order for human beings.[27] Stated another way, his work attempts to redefine the human condition by denaturing death. This is not a minor taxonomic leap, and it most certainly creates a new kind of ontology for the human or, at least, a new kind of ontological status for the human corpse. Death is not ontologically impossible in this scenario (it is actually necessary to produce the exhibitions), but the visible dead body has a new kind of immortal defiance of nature. This is another way of stating Virno's argument: if the dead bodies in *Body Worlds* labor hard enough, maybe no one will notice that their ontological potential is impossible. Spectators know that the plastinated bodies are dead (signs clearly state this fact at *Body Worlds* exhibitions), but the same viewers embrace these postmortem humans as somehow impossibly vital and dynamic.

This becomes the ultimate taxonomic power over nature: we humans, or at least our bodies, can live forever because we pull ourselves from nature. Von Hagens goes so far as to say that not even the ancient and much revered mummies accomplished this feat. He also clearly states that his understanding of the "denatured body" means a dead body that does not decompose. But this denaturing goes further than von Hagens intends. In this reinvention of nature, von Hagens is less a noble modernist than an eighteenth-century Linnaean taxonomist trying to redefine human nature by simply creating new rules for death, new forms of being, and new forms of labor for the dead body.

Whether von Hagens realizes it or not, his *Body Worlds* exhibitions succeed by using the language of anatomical science to further indulge man-made fictions. It is this relationship between science and fiction that comes closest to explaining how mythological creatures can exist alongside scientific displays of human remains. The Linnaean taxonomist needed the centaur to explain how the human is not a mythological creature, just as von Hagens needs the pathologization of anatomy to present postmortem sexuality as normal. Michel de Certeau described the science and fiction relationship this way: "Through a rather logical reversal, fiction may have the same position in the realm of science. ... The 'fiction' is not the photographing of the lunar space mission but what anticipated and organized it."[28] At some point in the past, when centaurs and mermaids stopped being real and died, human primates were left with a taxonomic gap that required a new human identity. Just as landing on the moon means humans can defy gravity (an impossible fiction in its own time), denaturing death and its affiliated ontologies means humans can in fact live forever. It is both a fiction and a possibility. Von Hagens's denaturing of death begins not with science but with mythology.

The Release Form

Question #7: Plastinated specimens, especially whole body plastinates, are occasionally interpreted as anatomical works of art. Hence the question: I agree that my body can be used for an anatomical work of art.[29]

—Institute for Plastination, *Body Donation Program* release forms

In recent years von Hagens stopped accepting requests from living individuals to donate their bodies for use in exhibitions—the Institute for Plastination simply has enough donors on the books and dead bodies in storage. But before any individual could previously donate their body to von Hagens, they first filled out the *Body Donation Program* release form. It was a multipage document that asked a series of questions about why people wanted their bodies used in the displays. Is it okay, the forms asked, for your body to be used for medical training? Would you like your body to remain anonymous? These are a few examples. The forms also stipulated the basic terms that all parties agreed upon once the donor signed the documents. One of the donor program's final clauses was rather important. It stated: "I agree that my preferences as indicated in the following questions are recommendations rather than binding terms. The IfP [Institute for Plastination] makes every effort to fulfill body donor wishes. Particularly with regard to younger body donors however, it cannot guarantee individual outcomes."[30] Once a body was donated for plastination, total control of that body rested with the IfP.[31] This donation clause was not particularly nefarious since it made good sense to state upfront that a donated body may or may not be used as the donor wished.

What this language truly revealed was how much control Gunther von Hagens exerted (and exerts) over the *Body Worlds* exhibitions. He was able to exert control over concepts of life, death, anatomy, pathology, taxonomy, ontology, mythology, science, fiction, and sexuality like few others. Once the paperwork was signed and the donor died, von Hagens owned that corpse. His posing of dead bodies in sexual positions or mixed with animal parts is simply the fullest exertion of that legal authority. This kind of total control is not necessarily bad, and in no way does von Hagens seem to espouse a malevolent agenda. In June 2008, for example, von Hagens met in Los Angeles with over one hundred people who agreed to donate their dead bodies to him. He heroically (if not a little self-servingly) described those individuals as gaining "'post-mortal' citizenship" in *Body Worlds*.[32]

Post-mortal citizenship notwithstanding, it is still fair to ask why? Why pose dead bodies in sexual positions? Why mix human and animal body parts together? The most honest and potentially oversimplistic answer is this: von Hagens will display dead bodies in these ways because he can. He is succeeding in ways that eighteenth-century taxonomists, early anatomical clinicians, and purveyors of "obscene" pathological displays could only dream about. The more complicated response to either question is this: Positioning dead bodies in sexual positions and creating mythological creatures is the logical conclusion to human struggles with biological taxonomy since the eighteenth century. It is also a further example of postmortem technological mediation that suggests even in death *Homo sapiens* can defy all animal kingdom conventions. And it satiates, as Gunther von Hagens suggests, "the longing to be immortalized."[33] Most importantly, it is a chance to be more than naturally human.

It is easy to comprehend how these exhibitions, such as 2009's *The Cycle of Life*, offend large numbers of people—a situation that only guarantees record-breaking attendance. But the people who take offense miss a larger point. *Body Worlds* viewers know that dead bodies cannot ride dead horses, jump hurdles, or shoot arrows—even though audiences willingly accept these narratives. An audience member's willing acceptance of these specific narratives betrays the following sentiment: dead bodies should only be productive in *certain kinds of socially acceptable, completely impossible ways*. Von Hagen's *Cycle of Life* exhibition does not display too much human anatomy; rather he poses that anatomy in a most self-aware, human act that uses the most impossibly unaware kind of human bodies. Ironically enough, and it is always worth stating, von Hagens needs offended people so that the moral qualms that surround his work can enable his exhibitions to function. In a most peculiar way, von Hagens has simultaneously reinvented the dead body for both curious audiences and museum administrators, irrefutable proof that if no one objected to von Hagens's use of dead body technologies, he would have disappeared long ago.

8/18/2018

Watching My Sister Die—10 Minutes/10 Days
I've got 10 minutes old friend
before I   start the day
10,000 feet   in the sky.
Going home to see my parents
to see them without my sister
who will never join us again.
As I become their only living child.
And I lost you old friend
for   72 hours
leaving you on the plane that brought me back
from my sister's funeral.
In those days I realized how important
everything in these 10 years of pages is to me.
my life
in ink
and so many tears.
So I promise to never   leave you behind again.
It's time to go   old friend.
To re-discover the past
And hold the unknown tight.

Holding you close too.
In the increasingly disjointed days of my mortality.

5   The Global Trade in Death, Dying, and Human Body Parts

> The vitality of a culture or ideology depends upon its ability to channel the power of such mordant symbols as the corpse.
>
> —Richard Huntington and Peter Metcalf, *Celebrations of Death* (211)

A quasi-invisible postmortem economy has taken shape since the latter part of the twentieth century as a direct result of new and ever expanding dead body technologies that humans began using in the late nineteenth century. It is an economy structured around the human corpse that is simultaneously global in scope and local in procurement. It is an economy that is actually quite visible to the human eye if a person knows where to look, but it is also an economy whose actors seek anonymity. This is the global trade in death, dying, and human body parts, and it represents a postmortem biomaterials industry that gladly helps families decide what to do with a dead body. Most next of kin can, of course, assign a social or familial value to a person's dead body—but what about a monetary value? It is a new twist on the questions raised in all the previous chapters: How much is a dead body worth? What is the current, everyday commercial value of dead bodies, body parts, and postmortem human tissues?

Martha W. Anderson and Renie Schapiro suggest that a whole dead body's average monetary value is $30,000 to $50,000. The financial value can also reach "to over $200,000 for donors from which the tissue is processed into medical implants, demineralized bone matrices, or dermal implants in addition to traditional bone and skin grafts."[1]

The focus here is primarily on postmortem tissues, bones, and body parts and does not delve that deeply into the illicit sale of human organs. In the United States, both the National Organ Transplant Act and the Uniform Anatomical Gift Act make it explicitly illegal to profit from the sale of organs, especially ones being used in human transplantation.[2] Focusing on postmortem tissues enables a closer examination of how biomaterials, which lack donated organs' noble social standing and legal protections, produce significant sums of money. How the laws governing the sale of organs and human tissues intertwine is discussed, but important sociocultural value distinctions still remain between heart transplants and cadaveric skins grafts. Dead bodies offer a wide range of postmortem biomaterial possibilities that become financially exploited long after surgeons remove all the transplantable organs.

Necroeconomies and the Dead Body

In *Tissue Economies: Blood, Organs, and Cell Lines in Late Capitalism*, Catherine Waldby and Robert Mitchell argue that "within the body, tissues constitute the biological substrate of the self, the condition of viable human life. Once donated, they can sustain the life and health of another. A tissue economy, in our terms, is a system for maximizing this productivity, through strategies of circulation, leverage, diversification, and recuperation."[3] What makes these tissue economies important is their production of human biovalue, a concept the authors explain "is the surplus of in vitro vitality produced by the biotechnical reformulation of living processes."[4] These two concepts, tissue economies and biovalue, lay the groundwork for assessing a living body's physiological commercial value, but these concepts also open a door to discussing postmortem commodity values. Two concepts related to, but distinct from, tissue economies and biovalue more accurately describe a dead body's potential commercial value: necroeconomies and necrovalue.

Reconfiguring Waldby and Mitchell's terms in this way does not simply replace words that suggest life for words that mean the dead body and/or connote death. On the contrary, the distinctions between these kinds of biomaterial commodities illustrate how human corpses both represent and yield far greater opportunities for body parts and tissue removal than do living bodies. Living human bodies naturally restrict the amount of safely removable organic material and offer finite biological possibilities. Dead bodies, conversely, do not require any biological materials for long-term functioning.

Given these considerations, and with a deep intellectual debt to Waldby and Mitchell, the following dead-body-specific concepts articulate how the global trade in death, dying, and human body parts functions: *Necrovalue is the yield of postmortem biological materials produced by the necrotechnical repurposing of the human corpse.* Necrotechnical repurposing is most easily defined as the combination of both practical and political technologies used on a corpse since the nineteenth century to radically alter its value. These combinations include everything from mechanical embalming, to biomedical products such as tissue implants, to legal categorizations of postmortem ownership. The HIV/AIDS corpse is yet another example where an entirely new kind of necrotechnical interface was invented to handle the dead body, but for different historical reasons related to the technologies of the dead self. In more provocative terms, Gunther von Hagens has done the same kinds of things with the plastinated corpses in *Body Worlds*.

Waldby and Mitchell go on to state that "tissue economies are at base about the way the biological capacities of the human body contribute to social, economic, and political systems of productivity and power."[5] The following distinctions accordingly emerge from their points: *Necroeconomies are fundamentally about how the postmortem biological potential of the human corpse, e.g., the dismembered dead body that produces multiple medical implants, etc., contributes to the "... social, economic, and political systems of productivity and power."*[6] A shortened version of this entire argument simply states that necrotechnologies are used on human corpses to create necrovalue, which then produces necroeconomies. All of these dead-body-specific technologies, economies, and values necessarily require the human corpse. A living body is no use to technologies built around stripping the dead body of its usable postmortem biological materials. Or, a living body could be potentially useful but only after it is made dead—an older corpse procurement strategy that William Burke and William Hare notoriously used for a time in nineteenth-century Edinburgh before authorities stopped them.[7]

During the first half of the 2000s, a headline-grabbing necro-economy court case (not entirely dissimilar to Burke and Hare's—minus the murders) demonstrated how the necrotechnical repurposing of the human corpse could turn the dead body into a financially lucrative, internationally distributed anatomical product. The case involved a former Manhattan dental surgeon,

Michael Mastromarino, and his company Biomedical Tissue Services (BTS). While the nineteenth-century American postmortem model embraced safely transportable, intact dead bodies for funerals, advances in twentieth-century biomedicine found new opportunities for the corpse's dismembered necrovalue. Mastromarino and Biomedical Tissue Services simply took the dead body's exponentially expanded commodity value to its logical, if not extreme, conclusion.

Necroeconomies at Work

In October 2006 *New York* magazine published one of the first, lengthy investigative reports on Michael Mastromarino and his involvement in the burgeoning, global postmortem body parts market. The article explained that "Mastromarino was once a prosperous maxillofacial surgeon, with offices in New Jersey and midtown Manhattan ... he was chiefly known as coauthor of *Smile: How Dental Implants Can Transform Your Life.*"[8] His contribution to that text, the article explains, was "the chapter on bone-grafting, which the book says is 'a whole new ball game.'"[9] Randall Patterson, the article's author, goes on to discuss how it is that Mastromarino became internationally well-known not for his dental surgeries but for a drug addiction problem that forced him to surrender his dental license and that eventually led him to buying and selling postmortem human body parts. Patterson describes how Mastromarino found his entryway into the necroeconomy through contacts at a large tissue bank, "a company that trades on nasdaq under the name RTIX and that manufactures ... all sorts of useful spare parts, including the 'BioSet Demineralized Bone Matrix (DBM),' the 'patented MD-Series threaded bone dowels,' 'Osteofil/Regenafil injectable bone paste,' as well as a number of 'cortical bone pins' and 'interference screws.'"[10]

It was through this convoluted path that Mastromarino created his own company, Biomedical Tissue Services, in order to harvest bone and tissue from human cadavers. He then sold the biomaterials to some of the world's largest human tissue and bone product producers. By putting his surgeon's skills to work, Mastromarino appeared to turn his life and financial fortunes around. The only problem Mastromarino faced was that he could not satiate the biomedical corporations' demand for tissues and bones. He began hiring employees, and fairly soon all manner of bodies were being dissected for bones and tissues. Here is where the trouble began for Biomedical Tissue Services and Mastromarino. As with so many situations involving an in-demand, financially profitable product, Mastromarino began to favor greed over following the law. He started forging signatures on next-of-kin consent forms and the related paperwork; multiple bones and tissues were taken from bodies where the cause of death was changed, and profitability overcame any concern about a dead body's suitability for use in biomedical products. In early 2006, Mastromarino's (then alleged) crimes began to publicly emerge, and he met with widespread, global condemnation. But even before his legal problems became public, it was clear that the situation was spiraling out of control.

On October 13, 2005, the US Food and Drug Administration (FDA) circulated a "Recall of Human Tissue" memorandum for Biomedical Tissue Services. The recall centered on unsafe human tissue and bone sold by BTS for use in medical procedures. The FDA memo stated: "Biomedical Tissue Services (BTS) was recently made aware that there is the possibility that tissue has been procured from donors without proper medical/social histories. BTS is performing a voluntary recall of any unused tissue from its consignees."[11] A few weeks later, the FDA published an update "to strongly recommend that health care providers inform their patients who receive tissue implants prepared from BTS donors that they may be at increased risk of communicable disease transmission and to offer them testing."[12] The recall subsequently affected postmortem biomaterials used in America, the United Kingdom, and Canada.[13]

Finally, on February 23, 2006, the Kings County District Attorney in Brooklyn, New York charged Michael Mastromarino and a small group of his employees with 122 counts of, but not limited to, "Enterprise Corruption, a Class-B Felony punishable by up to 25 years in prison, Body Stealing and Opening Graves (Class-E Felonies), Unlawful Dissection (an unclassified Misdemeanor), Forgery in the Second Degree, [and] Grand Larceny in the Third Degree (Class-D Felonies)."[14] The Kings County District Attorney summarized the indictments at a press conference: "[S]tealing tissue from the dead ... is like something out of a cheap horror movie. But, for the thousands of relatives of the deceased whose body parts were used for profit, and the recipients of the suspect parts, this was no bad movie. It was the real thing."[15]

One of the most commonly used examples of how Mastromarino and Biomedical Tissue Services harmed the deceased, next of kin, and living tissue recipients involved radio and television host Alistair Cooke. Cooke had been the host of Masterpiece Theatre on PBS for over twenty years and presenter of the BBC's *Letter from America* for over fifty years. In a March 2006 *New York Times* op-ed, Alistair Cooke's daughter, Susan Cooke Kittredge, recalled the phone call from a New York City police detective who informed her that her father had had tissues and bones removed by BTS: "Apparently investigators found that those who sold his tissue had falsified my father's age and cause of death, listing him as 85 rather than 95 and as having died of a heart attack rather than lung cancer that had metastasized to his bones."[16] The next-of-kin signatures on Cooke's tissue removal consent form were also forged before his body was dismembered and then cremated.

After further investigation, it emerged that Mastromarino had worked with funeral directors in three different states, New York, Pennsylvania, and New Jersey, and that he paid roughly $1,000.00 in cash for each corpse. Parts of the broader legal case also took seemingly absurdist turns when, for example, the Pennsylvania criminal trial required a different judge. Three of the Philadelphia funeral directors who provided tissues to Mastromarino also played golf with the already presiding judge, so a nongolfing jurist had to be assigned the case.[17] On March 18, 2008, Mastromarino pleaded guilty to the Kings County charges, agreed to pay the District Attorney's office $4.6 million dollars for operating a corrupt business, and was sentenced to eighteen to fifty-six years in prison.[18] His story took an unexpectedly tragic and somewhat karmic twist when, on July 7, 2013, Mastromarino died in prison from liver and bone cancer.

Incomprehensible as it might sound, the actual collecting of human tissues and body parts by Mastromarino and his conspirators was, by itself, not illegal. Mastromarino and his employees only broke the law when they began forging paperwork, lying about the provenance of the biological materials, and not acquiring next-of-kin consent for tissue harvesting. The legal infractions did not directly involve the biological materials, per se, but forgery, greed, and lying. Necroeconomies and the basic concept of using dead bodies for financial gain occupy a peculiar space within the law. The procurement system is deceptively straightforward: buyers and sellers never directly pay for the human body parts that they deliver or receive. These individuals pay or receive a *handling fee*. Michele Goodwin, in *Black Markets: The Supply and Demand of Body Parts*, explains: "Transactions in the tissue-processing industry are problematic, but seemingly protected by a loophole in the National Organ Transplantation Act, which provides for reasonable fees to be used in the transporting and processing of human body parts."[19] The FDA does regulate and oversee the sale of human tissue and bone, as demonstrated by the 2005 warnings about Biomedical Tissue Services, but none of the regulations cover the "reimbursement fee" or the "reasonable fee" that companies charge to handle postmortem body parts. Mastromarino and his employees never received direct payments for the postmortem tissues and bone; they collected millions of dollars in "processing fees." By making payments in this way, no tissues, bones, or intact cadavers were actually purchased; buyers only paid for time and labor. This is also why the Brooklyn District Attorney's office never pressed charges that included the actual biological materials.

Mastromarino and other individuals who work in the necro-economy sometimes refer to themselves as *body brokers*. Body brokers work as middlemen between buyers (e.g., biomedical corporations such as Johnson & Johnson) and the three usual sources of human cadavers: university hospitals with medical schools, funeral homes/crematoriums, and county morgues that perform autopsies.[20] In the event any of the financial rewards accumulated by body brokers collecting handling fees sounds implausible, Annie Cheney's 2006 *Body Brokers: Inside America's Underground Trade in Human Remains*, clearly documents how certain kinds of necroeconomies function. Her book helped reveal just how valuable necrotechnically repurposed postmortem biological materials can become. Prices for the handling of human body parts may vary from body broker to body broker, but Cheney documents "heads $550, brains $500, shoulders $431, spines $1500, knees $500, tibias $400, femurs $467.30, whole legs $815, feet $350, forearms $350, five grams of skin $803.57, vaginas (with clitoris) $350, breasts $375, and fingernails $15 a piece."[21]

Mastromarino's criminal activities were also not a new story. Much press coverage made historical connections with nineteenth-century grave robbers, body snatchers, or "ghouls."[22] But a person need not reach back that far, if for any reason, Mastromarino was simply one actor in an enormously populated necroeconomy of body brokers, biomedical product makers, and dead bodies. Two other postmortem "scandals" also occurred at roughly the same time as Mastromarino's: in the UCLA medical school and with an independent body broker named Philip Guyett. Whenever one of these body parts scandals emerges, the situation is often described as if it existed in its own unconnected, hermetically sealed vacuum. These cases are most certainly interconnected, not by direct person-to-person relationships but by the overarching financial gain represented by necroeconomies. The main reason that anyone in the general public knows about these postmortem tissue trafficking cases is because an individual got caught. Mastromarino and his fellow body brokers are usually only visible after the fact.

Henry Reid, Ernest Nelson, and the UCLA Medical School

In March 2004, Henry Reid, the overseer and central administrator of the University of California at Los Angeles's Willed

Body Program, was arrested for providing human cadavers to another man, Ernest Nelson, for the purpose of dissection. Mr. Nelson then sold the body parts removed from the cadavers, by his own admission, to "'giant' medical research companies."[23] In response to the arrests, the UCLA medical school temporarily suspended acceptance of any new bodies to its cadaver program and began a full investigation into how this trafficking in human body parts could have ever occurred. This very publicly announced investigation began even as Ernest Nelson repeatedly explained to reporters that "medical school officials gave him access to the university's body freezer twice a week, where he was allowed to saw off knees, hands, torsos, heads, and other body parts."[24] Goodwin gives the following details about the case as well as raising some broader necroeconomy issues: "Among Nelson's clients was the Fortune 500 pharmaceutical giant, Johnson & Johnson. Johnson & Johnson's subsidiary, Mitek, obtained tissue from Nelson in the 1990s ... UCLA is simply the canary in the coalmine. Other medical schools, university hospitals, and organ procurement organizations are known to engage in such clandestine transactions."[25] Johnson & Johnson quickly responded to the UCLA case. A spokesperson explained that while its subsidiary Mitek did in fact purchase postmortem biomaterials from Nelson, "Mitek did not knowingly receive samples that may have been obtained in an inappropriate way."[26] The spokesperson's response led to two persistent and unanswered follow-up questions: Where exactly did the company believe the postmortem body parts originated? And, did anyone ever ask?

After further investigation by authorities, both men were criminally charged in March 2007 with conspiracy and grand theft, and Mr. Nelson was charged with tax evasion on the over one million dollars that he made by selling human body parts to "more than 20 private medical, pharmaceutical and hospital research companies."[27] Reid pleaded guilty in October 2008 "to one count of conspiracy to commit grand theft, with a special accusation that he damaged or destroyed more than $1 million worth of school property, which refers to the donated bodies."[28] Finally, in May 2009, Ernest Nelson was found guilty of conspiring to commit grand theft, embezzlement, and tax evasion and in June 2009 sentenced to ten years in prison. He was also ordered to pay fines, penalties, restitution, and back taxes, which totaled more than $1.7 million.

Philip Guyett and Donor Referral Services

On August 18, 2006, the FDA hand delivered an "Order to Cease Manufacturing and to Retain HCT/Ps [human cells, tissues, and cellular and tissue-based products]" to Philip Guyett and his company Donor Referral Services in Raleigh, North Carolina. The FDA letter provided Guyett with six separate operational violations for his handling of eight different cadaveric tissue donors. The first of the six violations cites information Guyett recorded about a tissue donor in a report. Guyett wrote that the donor died at home, from a heart attack, did not use drugs, had no history of cancer, and was not in a care facility of any kind. The FDA later uncovered the donor's official death certificate, "which states that the donor died at ----- Nursing Center of rectal squamous cell cancer, with other significant contributing conditions being intravenous drug use, coronary artery disease, tobacco abuse and chronic obstructive pulmonary disease."[29]

Guyett had already run afoul of California law, where in 1999 he was charged with embezzlement for selling a dead body to a medical school and keeping the $1,100.00 for himself.[30] At that time Guyett was the Western University of California's willed body program administrator. After the most serious charges were dropped, Guyett was fined, served six months' community service doing highway cleanup, and received three years' probation. In 2003, Guyett registered Donor Referral Services with the FDA and began selling human tissues and body parts from Las Vegas. After one of Guyett's FedEx packages filled with cadaveric limbs began leaking while in transit to Missouri, police again investigated him, but nothing happened.

Then, in 2004, Guyett relocated Donor Referral Services to Raleigh, North Carolina, and began making his sales pitch to whoever would listen. As it turned out, Guyett found interested parties. A North Carolina funeral director, Larry Parker, explained "Guyett paid him $1,000 for each of the roughly 60 donors the funeral home referred. Parker, who began seeking donor cadavers for Guyett in 2004, said he believed other funeral homes also dealt with Guyett."[31] Guyett contacted a number of individuals and funeral industry businesses who in fact declined his offers, but that did not stop him from listing the names of those funeral homes as part of his business affiliates network. After pleading guilty to mail fraud, (i.e., how he handled and shipped the postmortem biological materials), Guyett was eventually sentenced in October 2009 to eight years in prison. Raleigh's *News and Observer* newspaper noted that "profit motive emerged as a factor behind a scheme which allowed potentially diseased body parts to be used in at least 127 unsuspecting patients around the country. ... Federal investigators found questionable records for another 40 donors, involving 2,600 human tissue products that went into the marketplace throughout the world."[32]

The Mastromarino-Nelson-Reid-Guyett Necroeconomic Model

The individuals in these three cases unintentionally succeeded in momentarily focusing public scrutiny on the global trafficking of postmortem body parts and tissues, as well as the broader necroeconomy built around dead bodies. Each of the participants, Reid and Nelson in particular, also produced a momentary panic across the United States in funeral homes, hospital morgues, health clinics, and medical school willed body programs with regard to (1) how dead bodies were being used outside the expected status quo and (2) that individuals financially profited from the illicit sales of dead body parts while simultaneously undermining the broader public trust in key social institutions.

Freelance dissectionist and autopsy technician Vidal Herrera provided the following comments on the UCLA case when the story first appeared: "Everybody knows who to call—the buyers, the sellers, the disarticulators, the schools, the crematoriums. It's a lucrative business."[33] Dr. Todd Olson, professor of anatomy at the Albert Einstein College of Medicine in New York, also commenting on the Nelson-Reid case, gave an even more illustrative description of the global trade in postmortem biomaterials and its lack of regulatory oversight: "A lot of money is changing hands. ... It is easier to bring a crate of heads into California than a crate of apples. If it's produce, authorities want to know all about it."[34] US Senator Charles Schumer brought forth legislation in response to the Mastromarino case, but industry groups lobbied against it. Schumer explained: "They said it wasn't needed. They said that 'Everything is under control,' but I had real doubts ... what we saw happen in the Brooklyn funeral home could well be happening in lots of other places both here and abroad, there's no real protection."[35]

Yet all three of these cases, each within a matter of years of or even overlapping with the others, represented not systemic aberrations but the everyday operations of a postmortem economy that requires dead bodies. For some body brokers, as well as the companies that purchase biomaterials, working within the necroeconomy simply means procuring dead bodies by whatever means necessary. These three legal cases also highlight an implicit point raised by Olson, namely, the potential public health concerns raised by the illicit trade in human tissues, bones, and body parts. These concerns are largely based around the lack of either state or federal law in the United States regarding the "repurposing" of human cadavers.[36] Repurposing dead bodies can involve taking a corpse donated for anatomical, educational dissection and then selling parts of that corpse to companies for biomedical products. The repurposing done by medical schools, as an example, is hard to track and rarely gets discussed except when a situation such as UCLA's becomes visible. On its own, the concept and practice of cadaver repurposing does not necessarily cross any dangerous ethical or moral lines. And it makes sense that if a medical school has a surplus of dead bodies, then it might look to see how those corpses could be used in other venues.[37]

Repurposing activities do not, however, always meet next-of-kin expectations. Even if a medical school's willed body program produces detailed waiver forms that clearly place ownership of the corpse with the institution, families still think of the deceased as their own. Donor families sued UCLA, but those cases largely focused on a breakdown in faith, trust, and dignity for the dead body, not ownership rights.[38] These legal suits also occurred even though the dead bodies involved were always going to be dismembered. The key difference, of course, is the dismemberment's original, *intended* educational purpose versus the sudden disclosure of postmortem profiteering. These particular familial ownership questions are hardly the worst-case scenario for institutions involved in necroeconomies. The worst-case scenario for biomedical products created from improperly handled dead bodies involves the possible transmission of disease.

Brian Lykins's tragic 2001 case brings these disease concerns to the fore.[39] Lykins (age twenty-three) lived in Minnesota and received a tissue implant from a cadaveric donor who shot himself to death in Utah. The suicide took place in early September of that same year. The anonymous, deceased donor's dead body went unrefrigerated for approximately twenty-four hours (which significantly enabled decomposition and bacterial growth) before Intermountain Donor Services in Utah sent the postmortem biological tissues to another company, CryoLife, in Georgia.[40] At that point, the removed tissues and bones were processed and turned into transplantable biomedical products. Roughly two months later, in November 2001, Brian Lykins received a bone chip implant for his knee. The surgery took place in a St. Cloud, Minnesota, hospital, and Lykins subsequently developed severe postsurgery complications. He was dead from an aggressive *Clostridium sordellii* infection within a week. Lykins's death was so unexpected and shocking that it caused the Minnesota state legislature to change state laws regarding what kinds of companies could provide transplantable tissues for surgical procedures. Minnesota will now only allow not-for-profit tissue banks to work in the state.[41] Regulatory actions, such as Minnesota's, are generally only taken after the fact and do not always produce widespread, substantive changes. Lykins died in 2001. The

Mastromarino, Guyett, and UCLA cases continued to evolve after his death, over a period of at least eight more years. And each of those cases produced lengthy legal suits also attached to regulatory concerns. Over three hundred tissue recipients attempted to sue tissue banks that received biomaterials from Mastromarino, but US District Judge William Martini dismissed those suits in October 2008. Judge Martini ruled that the plaintiffs could not show direct harm from the transplants.[42]

Future court cases will inevitably move forward, but as Jennifer Bittner (a person involved in a different but similar 2000 California crematorium case involving the selling of human body parts) matter-of-factly explained, "If you're cremated no one is ever going to know if you're missing your shoulders or knees or your head."[43] Bittner's statement might sound cynical, but she is also absolutely correct: the postmortem source for necroeconomies, that is, the corpse, is ultimately disposed of once the usable tissues are removed. Body brokers understand these particular dead body disposition scenarios all too well and rely on a general opacity for their work. Necroeconomies function and thrive because very few individuals actually understand how these same economies work. What the body brokers are helping to create, knowingly or not, is the next step for the technologies of the human corpse. This next step is more global in its reach, but no less local in its postmortem biomaterials procurement strategies. This next step also, and importantly, involves the American funeral industry.

The Global Trade in Death and Dying

Time and again, it is the funeral industry that finds itself involved in these necroeconomy cases. This is not to suggest

that duplicitous American funeral directors are in any way dou-
bling as body brokers. The Guyett case, for example, demonstrated that many North Carolina funeral directors and owners
of funeral homes had no interest in the financial offers. The
National Funeral Directors Association has also produced its own
Best Practices for Organ and Tissue Donation, which clearly pro-
hibit the worst kinds of body brokering.[44] Funeral directors and
funeral homes do find themselves increasingly on the receiving
end of tissue donation solicitations, so much so that the NFDA's
2011 Policy on Organ and Tissue Donation stated that while revised
best practices post-Mastromarino have yielded some changes to
problems such as troubling disclosure ethics, and the postdona-
tion physical condition of the corpse, other issues persist. The
problems "are increasing as tissue banks and other procurement
organizations become more aggressive in their marketing leading to, among other things, concerns over the use of funeral
homes as recovery sites and funeral directors actively involved
in the consent and referral process."[45]

The NFDA's institutional guidance makes sense since the
future use of dead bodies in biomedical products will increasingly
rely on the funeral industry, and for two distinct reasons. The
first reason is somewhat obvious: funeral directors and funeral
homes can reasonably access postmortem biomaterials. The second reason is less obvious but equally important: the American
funeral industry is itself becoming more globally consolidated
and likewise expanding in new economic directions—not totally
dissimilar to the nineteenth-century changes transcontinental
railways caused. This new(ish) kind of postmortem corporate
business model suggests a potentially different state of affairs
for the continually evolving global trade in death, dying, and
human body parts. These shifts in the broader necroeconomy
also began over thirty years ago.

During the late 1980s and throughout the 1990s, two multi-national funeral home corporations began acquiring thousands of funeral homes, cemeteries, and crematoria. Those two corporations, Service Corporation International (SCI) in the United States and the Loewen Group in Canada, spanned the globe with their holdings. Each company used a standard corporate model that acquired local funeral homes for inclusion in the larger ownership portfolio. In most cases, the acquired funeral home's name would not be changed, the workers would be retained, and even the previous owner might work as a manager. What both SCI and the Loewen Group usually changed were the items for sale, such as caskets, and the costs for those goods. The offerings often became more limited, and the costs generally increased. Jonathan Harr, writing in *Funeral Wars*, explains: "[Ray Loewen] needed the increased margin of profit to pay for the capital costs of acquisition. It was his practice to raise prices immediately by as much as 15 per cent in every newly-acquired funeral home. He called these price increases 'revenue enhancement.'"[46]

The height of market power for these funeral industry corporations began to recede in the early part of the 2000s, but up until that time both SCI and the Loewen Group amassed large business holdings. Here are some numbers that provide an insight into the scales of ownership: In 1999 SCI owned four thousand funeral homes in twenty different countries, five hundred cemeteries and two hundred crematories.[47] This meant that by 2000, SCI "owned 25 percent of the Australian industry, 14 percent of the British and 28 percent of the French."[48] Many of those foreign holdings have since been sold off, and in some instances back to the original owners. The Australian holdings, for example, were sold in 2001. The Loewen Group, in 1999, "owned more than 1100 funeral homes and over 400 cemeteries in Canada and the United States and thirty-two funeral homes in Great Britain."[49]

In 1995, Mississippi funeral director Jeremiah O'Keefe took the Loewen Group to court over breach of contract and unfair business practices. O'Keefe had a standing contractual agreement to sell funeral insurance through another Mississippi funeral home, Wright and Ferguson, which the Loewen Group purchased in 1990. Individuals purchase funeral insurance to cover their future funeral's cost. These financial instruments can also be referred to as preneed plans or prearrangement plans. The plans enable individuals to save money on funeral expenses by allowing consumers to pay current market rates at the time of purchase as opposed to future, and more than likely higher, prices.

The Loewen acquisition interfered with O'Keefe's preexisting contract to sell funeral insurance plans, and the case resulted in a jury trial. The jury decided the case in O'Keefe's favor, awarding him $500 million in both compensatory and punitive damages. The jury arrived at this amount even though the damages being sought only totaled $125 million.[50] The largesse of the award was itself shocking to both O'Keefe and Loewen and eventually resettled by both parties to include "$50 million in cash, one million shares of Loewen stock at a guaranteed price of $30 a share, and a promissory note, payable over 20 years, to the amount of $200 million."[51] The court case and other mitigating circumstances resulted in the Loewen Group going through bankruptcy reorganization in 2002 and renaming itself Alderwoods.

In November 2006, Service Corporation International acquired Alderwoods for $1.2 billion dollars. The total purchase amount combined both hundreds of millions in cash and corporate debt.

After the acquisition, SCI was projected to earn nearly $2.5 billion in annual revenues.[52] SCI's *2011 Annual Report* described the company's operational health: "We are North America's largest provider of deathcare products and services. ... At December 31, 2011, we operated 1,423 funeral service locations and 374 cemeteries (including 214 funeral service/cemetery combination locations) in North America, which are geographically diversified across 43 states, eight Canadian provinces, and the District of Columbia."[53]

In March 2004, Mary Roach, author of *Stiff: The Curious Lives of Human Cadavers*, wrote an editorial for the *New York Times* about the UCLA medical school case, illicit body parts markets, and the funeral industry. She made the following suggestions: "Perhaps a modest financial incentive for donating one's remains to medical research will inspire more people to do it. ... And if a reasonable cash outlay is all it takes to get 10,000 or 20,000 Americans over their aesthetic qualms and into my camp, then everybody wins. Everybody but the funeral industry, and it has been winning long enough."[54] Roach is correct that a financial incentive would indeed encourage many more Americans to donate their bodies for medical research, although the ethical quandaries that surround direct payments for donation remain complex.[55] That said, her last point regarding the funeral industry is not entirely correct. The funeral industry has not won much of anything with the current postmortem biomaterials collection system, other than bad press coverage. Roach glosses over, although not intentionally, the following key point: the increasingly globalized and diversified American funeral service industry's economic future lies in transforming the mostly funeralization side of the business into a much broader, human body parts and tissue distribution system.

Service Corporation International is a good example of this economic transition in motion and conveys this shift by referring to itself as a *deathcare provider*. This is a linguistic turn that significantly widens what exactly an SCI funeral home can do within the funeral industry. Some of these institutional transformations are already occurring, as local tissue banks actively work with funeral homes to find new postmortem donors. A February 2008 article in the NFDA's *The Director* highlighted how these necroeconomic shifts are already occurring: "Carolina Donor Services ... has hired Dorman Caudle as its first full-time funeral home liaison. Caudle will be responsible for assisting funeral homes with the organ and tissue donation process. ... [He] served as a licensed funeral director in Winston-Salem, North Carolina, for more than 14 years."[56]

It is not difficult to understand how relationships between funeral homes and tissue donor services organizations make good business sense. Carolina Donor Services, for example, "reimburses funeral homes for reasonable costs directly incurred as a result of the donation process. Any costs related to donation are not passed on to the donor family."[57] Another tissue bank, the not-for-profit LifeSource based in St. Paul, Minnesota, handles organ and tissue donation for Minnesota, North Dakota, South Dakota, and part of Western Wisconsin. LifeSource also reimburses funeral directors for their labor and offers the following fees: "Organ donors without autopsy: $100. Heart valve only donors without autopsy: $100 (No reimbursement will be offered when autopsy follows heart valve only donation). Tissue donors including bone, tendons, fascia, vessels and pericardium: $200. Surgical skin donors: $100."[58] The NFDA's own guidelines certainly support these kinds of partnerships. But if the goal in establishing these referral systems and partnerships is to help legitimize and moderately regulate necroeconomies, then this question still needs answering: How much money is a dead body worth? Annie Cheney notes, "Many tissue banks market their services heavily, sending representatives to hospitals, funeral homes, nursing homes, morgues, and hospices to entice families of corpses and corpses-in-waiting to donate ... they sometimes offer families a free cremation and reward funeral homes with a commission for each referral."[59] But a free cremation is nowhere near the amount of money biomaterials companies make with these human tissues.

The American funeral industry should, and seemingly will, begin to look more seriously at involvement in well-regulated, legitimate necroeconomies. The main reason is this: pursuing the human corpse's necrovalue is the logical progression for a whole system of postmortem technologies that require the dead body but do not necessarily need a funeral. Or, as with the HIV/AIDS corpse, the technologies must adapt to shifting postmortem conditions that alter the supposedly stable corpse and the very "self" attached to that dead body. What American funeral homes can do is eliminate the middlemen that work with tissue banks and deal directly with the biomedical product companies that require postmortem tissues. Funeral directors, and the American funeral industry writ large, can then implement stricter controls and best practices using the NFDA guidelines as a template. This will mean that biomedical product corporations conceivably pay more for postmortem biomaterials, but it also means reducing the chances of Mastromarino-esque litigation. Individuals such as Mastromarino and Guyett need funeral homes more than the funeral industry needs them.

In this new kind of necroeconomic relationship, funeral homes could place body donation directly into prearranged, preneed, or funeral insurance plans. As previously discussed, some individuals prepay for their funerals vis-à-vis these arrangements. By working an optional body donation into the prearrangement package, economically disadvantaged individuals and families (as a more than likely targeted group) could afford funeral services previously outside their price range. What must also happen, however, is that these new postmortem technological opportunities for the economically struggling do not become predatory.[60] Similar kinds of proposals and plans are also emerging in other parts of the world. In October 2011, the United Kingdom's Nuffield Council on Bioethics published a long awaited report entitled *Human Bodies: Donations for Medicine and Research*. The Nuffield report encompassed both organ and tissue donation. One of the report's key suggestions was that an individual's funeral expenses would be paid if that individual was signed up to be a donor on the National Health Service's Organ Donation Register and the organs and tissues were usable. A new donor pilot scheme would also focus on "what, if any, role family members should have in authorising the use of organs in such circumstances, and whether expenses should be covered if in fact the person's organs prove to be unsuitable for transplant."[61] Implementation of a pilot plan structured in this way means that families will not be unduly burdened by a questionable postmortem quid pro quo. The pilot plan could also work like a prepaid funeral plan, where a deceased individual's wishes would be known.

Still, the potential for abuses looms large at the very suggestion of funeral homes working directly with biomedical corporations. These concerns are especially true for a globalized funeral industry where the core resource, the corpse, could come from anywhere in the world.[62] Waldby and Mitchell quite astutely address these ethical considerations for tissue economies: "[T]he present biopolitical situation appears to be riven with inequity. ... The surplus profit and biovalue generated by commercial biotechnology innovation depend on the dispossession of donors, while the wealthy in the North secure their health and longevity at the expense of the bodies of the poor in the South."[63]

The potential for abuse, particularly along the global North-South divide, is extremely real but potentially managed via a fully transparent, industry-wide commitment to strictly regulated practices overseen by the FDA. The NFDA, as an institutional example, has already laid the groundwork for these practices. Body brokers thrive in the shadows of postmortem commerce, and the corporations that purchase tissues from brokers also benefit from a lack of transparency and regulation. Purchasing prices are kept low, and profit margins remain high—but the situation does not have to operate in this manner. The American funeral industry is one of the few already extant industries that can and should play a much bigger role in not only discussing postmortem tissue collection but also making the process safer. As opposed to watching the necroeconomy's actors cause problems, American funeral directors can step into the situation and do what they do best: demonstrate how human-driven technologies producing postmortem conditions, economies, and corpses entangle human mortality in everyday life.

To suggest that necroeconomies are not messy, do not involve death, or do not require dead bodies misses the point. A necroeconomy can only exist by using the human corpse. As a result of this postmortem reality, we *Homo sapiens* stand on the edge of a possible new future. Just as nineteenth-century railways transported the intact human corpse to funerals for waiting next of kin, now global shipping companies send dismembered, repurposed postmortem biomaterials across the globe to help other, unrelated, human bodies live. This is simultaneously the reinvention of an old concept and an evolutionary change in scale. Neither of these points, it should be noted, guarantees *Homo sapiens* any escape from mortality.

8/18/2018

Watching My Sister Die—Gate 11
The tears started again, little sister
As soon as I walked toward the Departure Gates
Gate 11
Where I was sitting when you died

on the phone
only hours from holding your dead hand
so I took a photo of your Gate
touched it with my hand
And cried because I'm flying home now
where you will never be again.
In the rain
Always it seems in the rain.
This is harder than I expected, little sister.
Feeling you die again
At 10,000 feet.
So I'm crying and I don't really care what people think
     staring at the giant taking photos of Departure Gate
signs.
Reliving the day when the plane couldn't fly
fast enough.
The seat next to me is empty
And you'll always be sitting in it, Julie.
with me no matter where I go.

6 Biopolitics, Thanatopolitics, and Necropolitics

A free man thinks of nothing less than of death, and his wisdom is a meditation, not on death, but on life.

—Benedict de Spinoza, *Ethics* Book IV, Proposition 67

Human beings will confront a choice in the not-so-distant future about what death both offers and mandates for postmortem technologies that alter and shape the dead body. The Bisga Man, the living dead person on life-support machinery, the HIV/AIDS corpse, the plastinated dead body, and the necroeconomics of body parts all illustrate how living humans can exert both technical and political control over postmortem conditions. These mostly invisible technologies of the corpse also directly involve three distinct yet interconnected political concepts that explicitly shape the human concept of the human: biopolitics (a politics of life), thanatopolitics (a politics of death), and necropolitics (a politics of the dead body).[1] Each of these political concepts appears and often overlaps in any number of postmortem conditions when the politics of life, death, and the dead body collide. The management of HIV/AIDS and the HIV/AIDS corpse was one such example. Carl Lewis Barnes and Gunther von Hagens are two further examples of individuals working in distinctly different time periods, but who still exercised (and exercise) control over the dead body in ways that the other would recognize.

Given these political considerations, it is important to ask this question: What distinctions need to be drawn between biopolitics, thanatopolitics, and necropolitics, in order to recognize their political effects on both extant and emerging technologies of the corpse? A further question is this: What is the relationship of death and the dead body to the sovereign control (i.e., the state, local authorities, national politicians, religious denominations, etc.) of life? Answering these questions means momentarily, if not entirely, abandoning the historically constituted definition of *Homo sapiens*—an action made possible by contemporary human attempts at technologically controlling both death and the dead body. These human actions involve both authorities at the highest levels of national power and lone individuals seeking to control death's material reality. In other words, the full scale political production of dead human bodies without death, but also human bodies that resist becoming dead.

Dead Bodies without Death

Life and *death* are two of the keywords that make dead body technologies tangible. Each of these terms produces different kinds of political meanings and uses that then explicitly shape a third keyword, the *corpse*. As Giorgio Agamben explains, the Ancient Greek words for *life* were in fact two different terms and produced two different meanings. The Ancient Greeks "used two semantically and morphologically distinct terms: *zoe*, which expressed the simple fact of living common to all living beings (animals, humans, or gods) and *bios*, which signified the form or manner of living peculiar to a single individual or group."[2]

The bios, for Agamben, is a form-of-life, and the zoe is the concept of living as a condition common to all beings. Agamben then develops these distinctions about the concept of life in explicating the differences between "naked life" and "form-of-life." He lays out parameters for the zoe and the bios: "By the term *form-of-life* ... I mean a life that can never be separated from its form, a life in which it is never possible to isolate something such as naked life."[3] The form-of-life, as different than naked life, can take on many qualifiers to define a human and represent the modes in which life becomes lived. As Agamben suggests, "[T]he voter, the worker, the journalist, the student, but also the HIV-positive, the transvestite, the porno star, the elderly, the parent, the woman."[4]

These two distinctions about the word *life* in biopolitics are useful since similar points should be raised regarding the use of the word *death* in thanatopolitics. In Ancient Greek, death is translated into different concepts but remains today in two distinct forms: *nekros* and *thanatos*.[5] Thanatos is death as a concept, as well as the Greek God of death when used as a proper name. Nekros is a corpse and suggests the dead body in all forms.[6] It is correct to say that both necro and thanato mean "death," but each word's use connotes quite specific conditions. As with zoe and bios, thanatos is death that affects all being and nekros is a form-of-death, as in the human corpse. Nekros defines a kind of body while thanatos surrounds the body as the immanent possibility of death. If a sovereign authority's biopolitics, per Agamben's suggestion, focuses on managing life and its thanatopolitics encompasses managing death, then a necropolitics manages dead bodies.[7] Stated another way, necropolitics is the political power most explicitly related to dead body technologies, and it only produces, manages, and encompasses human corpses. The necropolitical has little ultimate use for either "life" or "death," even though these terms intersect over time.

Biopolitics, on the other hand, requires the constant possibility of death as its equal part nemesis and collaborator in maintaining social order. Michel Foucault makes this very point in *The History of Sexuality, Volume I*, by focusing on the classical sovereign king's right "to *take* life or *let* live."[8] The threat of death, but more importantly the juridical ability to force or cause dying, remains the ultimate power of the modern state. Necropolitics and thanatopolitics, conversely, do not require living for their persistence. To make a person dead means concerns over life are secondary at best. So, for example, a human corpse is not threatened by the prospect of forced living.[9]

In Foucault's 1975–1976 *Society Must Be Defended* lectures, where his concept of biopolitics first appears, he outlines and argues that since modern-day state authorities now almost exclusively intervene to make individuals live (through public health campaigns, reduction of risky behaviors, etc.), death becomes the end point of that same power. Where the sovereign king once *made die and let live* (following the Foucaultian analysis), now the state *made live and let die*—but only to a point. Sovereign power may infiltrate and manage life at every turn, but it still remains unable to exert total control over death. He explains: "Death is beyond the reach of power, and power has a grip on it only in general, overall, or statistical terms. Power has no control over death, but it can control mortality. And to that extent, it is only natural that death should now be privatized, and should become the most private thing of all. ... Power no longer recognizes death. Power literally ignores death."[10] For Foucault, an individual's mortality now represents the simultaneous possibility of both living and dying in any given moment; it functions as the midway point between the sovereign's attempted control of both life and death.

Implicit in Foucault's argument that power "ignores death" is the suggestion that sovereign power indirectly acknowledges the impossibility of controlling death by turning a blind eye toward it. Any and all things must be done to keep a person alive; death will be fought until the limit is reached.[11] The act of ignoring, however, suggests that whatever is being ignored is, or has necessarily been, recognized as present by authorities. But by ignoring death, sovereign authorities hardly make the end of life or dead bodies disappear. On the contrary, what ignoring death makes inadvertently visible are the technologies that then support the power struggle over mortality between the individual and the state. Dead bodies often emerge from this same struggle, but the human corpse is also biology's challenge to sovereign power's limited control over death.

Agamben uses the WWII German internment camp system in *Remnants of Auschwitz* as a clear historical example that further explicates the challenge posed by death to the control of life. He explains that in the camps "an unprecedented absolutization of the biopower to *make live*," intersected "... with an equally absolute generalization of the sovereign power to *make die*, such that biopolitics coincid[ed] with thanatopolitics."[12] For Agamben, the collision between biopolitics and thanatopolitics in the internment camps suggests a paradox, i.e., that an absolute power to make live and an absolute power to make die cannot simultaneously coexist. When Agamben discusses the biopolitics and the thanatopolitics of the camp, he is also, arguably, discussing the production of dead bodies. The seemingly

impossible, simultaneous overlap between biopolitics and thanatopolitics in the camps was then negotiated by the explicit
emergence of necropolitics; necropolitical technologies transformed the human body from a living state to a dead state without any acknowledgement of death or dying taking place. Death
(in any individualistic or non-industrial-scale production sense
of the word) was totally ignored. Necropolitics facilitates this
production of *dead bodies without death* through the radical rejection of death from both biopolitics and thanatopolitics. Agamben describes what the camps accomplished: "In Auschwitz,
people did not die; rather, corpses were produced. Corpses
without death, non-humans whose decease is debased into a
matter of serial production. And, according to a possible and
widespread interpretation, precisely this degradation of death
constitutes the specific offense of Auschwitz, the proper name of
its horror."[13]

These bio-thanato-necro relationships, especially in the
internment camp context, involve postmortem politics and
technologies of the corpse that extend well beyond the Nazi system and into the present day. The danger in solely discussing the
camp as part of the Nazi regime is that these internment spaces
become overly fixed in that particular historiography. As a new
kind of political space constructed by authorities to control the
population, the *camp* represents for Agamben "the inaugural site
of modernity: it is the first space in which public and private
events, political life and biological life, become rigorously indis-
tinguishable."[14] In Europe and the United States (especially since
2001) the camp becomes the contemporary site for modern life,
death, and dead bodies to emerge as explicitly managed biological forces.[15] One of the single most important tools for these
camps is the language used to explain how prisoners went from

living to dead. The US government's prisoner of war or enemy combatant camps in Afghanistan, Iraq, and Cuba illustrate this very point regarding authorities' management of death and dead bodies.

These contemporary camps have led to the production of dead bodies where the true cause of death was simultaneously ignored and intentionally hidden. In many situations, an individual died as a direct result of torture and abuse but the camp authorities ignored the death by defining it as a natural, biological event. The American detention camps must rhetorically invoke biopolitics for interrogation purposes (i.e., to make the prisoners live) and not invoke necropolitics, since the publicly stated, governmental purpose for detaining individuals is to protect the nation precisely through the detention and interrogations—not to produce dead bodies. In 2004, Dr. Steven H. Miles from the Center for Bioethics at the University of Minnesota published an important article in *The Lancet* that detailed how the interrogation practices producing corpses in these camps were in fact hidden to make the deaths appear natural. Miles documents how:

A medic inserted an intravenous catheter into the corpse of a detainee who died under torture in order to create evidence that he was alive at the hospital. ... Death certificates of detainees in Afghanistan and Iraq were falsified or their release or completion was delayed for months. Medical investigators either failed to investigate unexpected deaths of detainees in Iraq and Afghanistan or performed cursory evaluations and physicians routinely attributed detainee deaths on death certificates to heart attacks, heat stroke, or natural causes without noting the unnatural aetiology of the death. In one example, soldiers tied a beaten detainee to the top of his cell door and gagged him. The death certificate indicated that he died of "natural causes ... during his sleep."[16]

These examples illustrate that when the authorities oversee-
ing a camp decide that a dead body is not a living body *made
dead* by the authorities in charge, rather a dead body produced
by an always-possible "natural death," then no wrongful act
ever occurred. Yet these same authorities still face a persistent
political problem: the presence of the dead body. It is the human
corpse, produced either naturally or unnaturally, that momentarily exposes the sovereign's inability to ignore, hide, and control death. The dead body remains irrefutable, political proof
that an individual died. To surmount this dilemma, authorities
simply stated that living bodies became dead bodies without any
recognizable form-of-death ever occurring. This impossible suggestion becomes possible when the biopolitical power to *make
live* and the thanatopolitical power to *make die* no longer matter. Necropower then takes hold as the unconditional sovereign
power *to make the living body dead without death* by totally ignoring the possibility of individual or biological death ever occurring. Agamben describes how contemporary sovereign power
came to so easily regulate and define human mortality: "This
means that today ... life and death are not properly scientific
concepts but rather political concepts, which as such acquire a
political meaning."[17] Death as both a concept and a reality of
biological life ceases to even register as a concern.

Bodies Resisting Death

What will happen to *Homo sapiens*' taxonomic status as the
twenty-first century unfolds is a curious situation that under-
scores the value of discussing the technologies of the corpse
alongside these bio-thanato-necro power relationships made
so visible by the camp system. While the camp system clearly

demonstrates how the power relationships between life, death, and the dead body can function, new kinds of mortality fights are beginning to emerge. Previous conflicts between the individual and the state over the management of a person's life that took shape in the 1970s will soon become secondary to twenty-first-century political contests over controlling death and the dead body. Not controlling death to attain an immortal life, rather controlling death to make it something different, something less absolute. The possibility of humans technologically controlling death and the corpse, however, is about far more than exerting just political control. Control of death in all its forms, by any authority of any scale and through every means possible, is also the largest remaining obstacle to redefining the "human" as a concept. That new definition of the human necessarily requires both the biopolitical control of life, the thanatopolitical control of death, and the necropolitical control of the dead body.

It is this ongoing political transformation and redefinition of dead body technologies into *death prevention technologies* (e.g., the necrotechnical revitalization of supposedly dead tissues for transplant) that suggests the control over human mortality is viable. If technologically controlling the dead body is possible, so one argument goes, then why not take a step back in time and physically eradicate dying from ever producing that human corpse. For this life-extension-merging-into-death-prevention concept to work, however, the radical, physical alteration of human mortality also requires an equally significant redefinition of how dead body technologies coincide with political power. These alterations to postmortem definitions, technologies, and politics are well under way, but so too are critiques of their possible taxonomic outcomes.

The technological control of the First World populations' mortality is being redefined and made increasingly possible as the entire human genome is broken apart and refolded into new combinations of organic biology merged with machines.[18] Gilles Deleuze suggested over thirty years ago that these genetic manipulations would produce what he called the *superfold,* something "borne out of the foldings proper to the chains of genetic code, and the potential of silicon in third-generation machines, as well as by the contours of a sentence in modern literature, when literature 'merely turns back on itself in an endless reflexivity.'"[19] The superfold comes at the end of Deleuze's book on Michel Foucault in an essay, appropriately titled "Appendix: On the Death of Man and Superman." A superfold becomes the Superman of Nietzschean description when "forces from within man enter into a relation with forces from the outside, those of silicon which supersedes carbon, or genetic components which supersede the organism ... in each case we must study the operation of the superfold, of which the 'double helix' is the best known example."[20] Deleuze suggests that the superfold is the production site of a new kind of human, wherein the mergers of external and internal forces constructing this new kind of being are themselves unstoppable and largely outside public view.[21] Deleuze cautions us about these changes: "As Foucault would say, the superman is much less than the disappearance of living men, and much more than a change of concept: it is the advent of a new form that is neither God nor man and which, it is hoped, will not prove worse that its two previous forms."[22]

What the superfolded human could potentially become, as a concept and increasing biological reality, is a new kind of taxonomic animal that escapes death and the dead body by technologically stopping physiological deterioration. As an example, groups of people have already formed to produce what the President's Council on Bioethics[23] called "Ageless Bodies" in its report *Beyond Therapy: Biotechnology and the Pursuit of Happiness.*[24] One of these groups, the World Transhumanist Association, seeks public understanding and acceptance of its core mission: "Humanity stands to be profoundly affected by science and technology in the future. We envision the possibility of broadening human potential by overcoming aging, cognitive shortcomings, involuntary suffering, and our confinement to planet Earth."[25] More recent writing about Transhumanist projects, especially Mark O'Connell's 2017 book *To Be a Machine: Adventures among Cyborgs, Utopians, Hackers, and the Futurists Solving the Modest Problem of Death*, document the "deathist" ideology (i.e., a belief system that makes the terror of dying seem acceptable) challenged by Transhumanist beliefs.[26]

It is worth noting that groups such as the Transhumanists (and the similar Posthumanists) are supported both financially and philosophically by a number of computer technology innovators now focusing on significantly extending human mortality through new kinds of computing machines. Ray Kurzweil, a prominent computer technology researcher and Posthumanist advocate, describes the manipulation of his own body in the following manner: "Genes are sequential programs. ... We are learning how to manipulate the programs inside us, the software of life. And personally, I really believe that what I'm doing is reprogramming my biochemistry."[27] What these ageless bodies portend is not a persistent physical adolescence but the prolongation of living by slowing down and outright stopping the body's deterioration over time.

And while contemporary online digital technologies and social media platforms exponentially increased during the

2000s, producing ageless human lives in dramatic new ways that theoretically challenged if people really ever died—these developments are temporary. Sooner than later, humans will again change what the internet does, and many dead people will finally disappear. It is also in that same moment of technological change that social media companies will likely become the world's largest online cemeteries, and I seriously doubt that company executives fully grasp how politically contentious graveyard decommissioning becomes.[28]

So one of the more important questions to consider is How might technologically augmented ageless bodies actually become "deathless" bodies? Or, potentially, death-resistant bodies that age at an exceptionally slow pace but still encounter decades-long physical deterioration? The President's Council on Bioethics report addressed these concerns by highlighting human mortality as a primary "Ethical Issue" given the steep overall decline in "untimely deaths." Yet, through the advent of modern antibiotics, life-support technology, and preventive medicine, a "stretched rubber band" life span might actually emerge. The report notes: "Under such circumstances, death might come to seem a blessing. And in the absence of fatal illnesses to end the misery, pressures for euthanasia and assisted suicide might mount."[29] The President's Council on Bioethics underscored, perhaps unintentionally, a primary biological concern with any human life extension: dying often becomes preferable to living at all costs.

Current laws that attempt to limit an individual from choosing to die may in fact become obsolete as suicide's current definitions simply become a new kind of normalized death, minus the associated legal or social repercussions. What biopolitical threats can authorities make when the durability of the human body is exponentially prolonged and death ceases to be undesirable? In the event that ageless bodies do become prevalent, an individual's choice to not supplement or change his or her body could itself be considered choosing a form of death. The question then becomes how much choice an individual will be given as a citizen of a nation and subject to its laws regarding how long to live or what to do with his or her body to prevent aging. Or, could permanent living itself become a new kind of punishment from the state? Not a lifetime sentence in a prison but a person *made to live* when in fact that body would otherwise die.[30]

The underlying technological and political issue for individuals, groups, and state authorities attempting to control mortality and ignore death is a seemingly overlooked but all the more fundamental point about human biology. George Canguilhem succinctly explained this phenomenon: "Life tries to win against death in all the senses of the word to win, foremost in the sense of winning is gambling. Life gambles against growing entropy."[31] Yet the human gamble with death means that both the individual and the state will make choices in the hopes of beating the biological odds against an inevitably dead body. If, and when, *total control* of death and dead bodies becomes the normalized power of human individuals or state authorities (a significant bio-thanato-necro power struggle unto itself), then the "human being" as a category of animal life will have crossed a threshold into a whole new state of existence. Deleuze's "death of Man" might actually become the successful rejection of human death through the total control of a humanly understood biological mortality.

Death has always been the human body's final act against its biology being pushed too far. The human corpse, as the end result of that action, is the one body that reminds individuals to think about death—if only in the simplest of terms. Total control of death, its administration and undoing, suggests not the end of dying but certainly the end of the human. In *The Open,* Agamben evokes Heidegger to suggest two potential outcomes for the future human: "(a) posthistorical man no longer preserves his own animality as undisclosable, but rather seeks to take it on and govern it by means of technology; (b) man, the shepherd of being, appropriates his own concealedness, his own animality, which neither remains hidden nor is made an object of mastery, but is thought as such, as pure abandonment."[32]

It is clear that we humans are in the midst of this first scenario as the technologies of the human corpse begin encroaching on death and working toward the prolongation of mortality. This fundamental shift in the necro-thanato-bio relationship suggests that at some point in the not-so-distant future the technological control of human life will have run its course. Humans will then necessarily abandon the concept of the "human" in the very moment that the second scenario comes into being. In the simplest of terms, abandoning the human necessarily means questioning the most fundamental assumptions about what makes the historically and mortally defined human being both a living and a dead body. Those questions will scrutinize the technologies of dead bodies as death becomes embraced rather than hidden within the animality of the human. It would seem that one day "pure abandonment" will describe a moment in the life of the human wherein *Homo sapiens* are no longer governed by a politics that *makes live, makes die,* or *makes dead bodies.* In that moment, none of these concepts will apply.

09/08/2018

Watching My Sister Die → Airports

I've been in Airports all summer, little sister
So many trips, I don't remember all the times
or places
But I remember where I was when I got the
call
the one that said you'd died.
And I'm not crying as much
but I'm still sad
The time is bending
it feels both like yesterday and centuries ago
when   I held your dead hand.
And said goodbye, one final time.
I feel like I'm running out of ways to talk about
you
little sister
And it's partly because your lived story ends at 43.
But we'll always have airports.

7   Patenting Death

Something is eating me.

I smoke too much
I drink too much

I am dying too slowly

—German playwright Heiner Müller in *Tooth Decay in Paris*, 1981

Whoever invents or discovers any new and useful process, machine, manufacture, or composition of matter, or any new and useful improvement thereof, may obtain a patent therefor, subject to the conditions and requirements of this title.

—Title 35 United States Code 101 regarding patentable inventions

On Friday, February 11, 2005, US Patent Application No. 20030079240—an application persistently rejected by the US Patent and Trademark Office (PTO)—was denied for the last time. Dr. Stewart Newman, from New York Medical College, originally filed patent application No. 20030079240 in December 1997 under the following descriptive title: *Chimeric Embryos and Animals Containing Human Cells*. The invention described in the patent sought to create:

A mammalian embryo developed from a mixture of embryo cells, embryo cells and embryonic stem cells, or embryonic stem cells exclusively, in which at least one of the cells is derived from a human embryo, a human embryonic stem cell line, or any other type of human cell, and any cell line, developed embryo, or animal derived from such an embryo.[1]

The *Washington Post* explained the invention in more general terms with an emphasis on the patent's anticipated end results:

Newman's application ... described a technique for combining human embryo cells with cells from the embryo of a monkey, ape, or other animal to create a blend of the two—what scientists call a chimera ... Newman's human-animal chimeras would have greater utility in medicine, for drug and toxicity testing and perhaps as sources of organs for transplantation into people.[2]

In a counterintuitive twist, Dr. Newman welcomed the final rejection not as a defeat but as an important bioethical victory. He wanted the PTO to either completely dismiss the original patent application or exclusively grant the chimera technology to him. Had Newman been granted the sole patent, he could have prevented others from gaining control over similar inventions for at least twenty years. By refusing to grant the patent to anyone, however, the PTO effectively blocked other parties from turning human-animal hybrids into patentable inventions.

Patent officials provided a handful of reasons for denying the application. The chimeras would violate the US constitutional right to privacy, officials suggested, since Dr. Newman or anyone he licensed the technology to would own the bodies in question as property. This potential property ownership could also violate the Thirteenth Amendment to the US Constitution. Whoever "owned" the patent also effectively "owned" the chimera made with the technology, and this ownership could itself become a form of slavery.[3] The application's final rejection was therefore the end result that Dr. Newman originally desired in 1997: the denial of a patent that would have legally recognized the creation and ownership of part human-part animal life forms. Dr. Newman explained his rationale for pursuing the patent's ultimate demise: "[T]he whole privatization of the biological world has to be looked at, so we don't suddenly all find ourselves in the position of saying, 'How did we get here? Everything is owned.'"[4]

What makes Newman's patent application case both fascinating and relevant is that similar biotechnical inventions, the very ones critiqued by Deleuze's superfold, will inevitably change how future concepts of death and the dead body represent the end of a "natural" human life.[5] The preceding pages discussed multiple situations and examples where controlling dead bodies undermined the supposed conceptual rigidity of death. In each case, from the Bisga Man to the Happy Death Movement to the HIV/AIDS corpse, from the denatured corpses in *Body Worlds* to the global trade in human body parts, to US authorities' necropolitical control in Guantanamo Bay, the conceptual stability of the dead body remained under constant reinvention. These very same postmortem technologies, and the productive results of those technologies, help to demonstrate how the logic of death that living humans believe to be so fixed is actually quite malleable. Herein sits the fundamental critique of the countless human technologies altering death and transforming the human corpse. The sum effect of how humans used and continue to use these technologies suggests that it is only a matter of time before the postmortem conditions that made the dead body look alive during the nineteenth century will move toward keeping the living twenty-first-century body from ever becoming dead.

Much of the scientific, legal, and ethical discourse surrounding Newman's application comes from an earlier, also applicable, patent dispute that concerned the ownership of manmade microorganisms—the 1981 *Diamond vs. Chakrabarty* US Supreme Court case.[6] In a landmark 5–4 decision, the Supreme Court found that human-invented living organisms could in fact be patented. The case involved a microorganism invented by General Electric Company (GE) engineer Ananda Chakrabarty that could digest oil from oil spills and industrial accidents. When Chakrabarty and GE filed the initial application in 1971, the PTO immediately rejected it. The application then worked its way through the court system until it reached the Supreme Court.

The reason that the Chakrabarty case is so important for both biotechnology research and technologies controlling the human corpse is that the court's decision made "life" a formally pantentable entity. Biotechnology critic, and coauthor of the Newman application, Jeremy Rifkin explains that "the PTO office rejected the [Chakrabarty] patent request, arguing that living things are not patentable under US patent laws."[7] And the majority decision in *Diamond vs. Chakrabarty* did in fact clearly state that: "The laws of nature, physical phenomena, and abstract ideas have been held not patentable. ... Thus a new mineral discovered in the earth or a new plant found in the wild is not patentable subject matter. Likewise, Einstein could not patent his celebrated $E = mc^2$; nor could Newton have patented the law of gravity."[8] Yet since Ananda Chakrabarty produced the microorganisms himself, he was *making* something new and not *discovering* something new. This distinction then allowed the Supreme Court to rule as follows: "A live human-made micro-organism is patentable subject matter ... [Chakrabarty's] micro-organism constitutes a 'manufacture' or 'composition of matter.'"[9]

The logic used in the Chakrabarty case, Newman's concerns regarding the private ownership of biological life, and the merger of both these cases with the constant evolution of dead body technologies bring me to this hypothetical question: What happens when human death is patented?

Death, it would seem, is a force of nature not unlike gravity, so any initial patent on "death" might automatically fail given the rules governing such applications. To legitimately patent death would mean more than simply inventing some new device or machine. What such a patent entails is finding a method, a concept, some combination of practical, political, and theoretical technologies that turn death into a man-made invention. Turning death into a humanly invented process would necessarily mean possessing both the powers to prevent it from happening, as is the case with current biomedical technology, and reversing previously irreversible death through this patented invention. By using the Chakrabarty case's reasoning, patenting death would also then mean its cessation was a "useful process" as defined by Title 35 of the US Code on patentable inventions.[10] Death would need to become entirely unnatural, both politically and practically, for the logic of the patent to succeed.

Any patent application for an invention that controlled death would necessarily need to stress how it assisted human mortality and to such a degree that the concept and practice of "life span" ceased to matter. Biomedical innovations that limited the finality of death were already visible in the latter part of the twentieth century as new forms of mechanical enhancement began keeping patients alive for increasing periods of time. As Dr. Robert H. Blank commented at that time, "Our very concept of what it means to be human is challenged by these rapid advances in medical technology."[11]

These conceptual and physical changes to the biological longevity of the human body are not, as this study of human corpse technologies suggests, without progenitors. The nineteenth-century preservation technologies that invented the modern human corpse provided a fundamentally important example of how these temporal shifts can operate. That new kind of nineteenth-century body offered the first moment in what later became the American public's desire to mediate and control death with technological precision. The contemporary medical and technological ability to replace malfunctioning human body parts with cadaveric tissues is a further demonstration of fundamental shifts in human biology made possible by corpses. These transplantation technologies certainly increased the possibilities for prolonging human life, but they also began bringing the immanence of death nearer its own limit. Agamben argues that the earliest stages of total human control over death appear through these persistent advances in human transplant surgery, beginning with the heart and eventually working to the brain: "According to any good logic, this would imply that just as heart failure no longer furnishes a valid criterion for death once life-support technology and transplantation are discovered, so brain death would, hypothetically speaking, cease to be death on the day on which the first brain transplant were performed."[12]

Agamben's point requires a slight addendum that the scientific question of whether brain transplantation is possible or not seems less an issue of feasibility than time.[13] How quickly the technology can be developed for safe use on humans is a more important question to ponder. The notion of a brain transplant also makes apparent that a person who lives without dying need not remain in the same body. Perhaps the person who dies from cardiopulmonary problems simply has his brain moved into another body, assuming another body has either been grown in a laboratory for such a purpose or is available from a donor. What these hypothetical technologies suggest is that the continued merger of humans and machines, as begun full force with the human corpse over a century ago, will push the human concept of the human into what can only be described as the terrain of science fiction.[14] As things stand today, the uploading of the dead self remains perpetually elusive (the possibility is somehow always about twenty years away), and companies that offer these digital postmortem preservation services allegedly require customers to commit "voluntary euthanasia" in order to access the brain.[15] *Caveat emptor* is my strongest advice. But as Dr. James J. Hughes, author of *Citizen Cyborg: Why Democratic Societies Must Respond to the Redesigned Human of the Future*, suggests: "Ultimately, the nanotechnological neuro-prosthetics that we develop to remediate brain injuries will also lend themselves to the sharing and backing up of memories, thoughts and personalities. That point may be recognized as the 'death of death.'"[16]

This idea, the "death of death," suggests that the ongoing fusion of living bodies with life-span-extending/death-preventing technologies found an important precursor in the corpse. As the machines making dead bodies look alive became the technological precursors of devices keeping almost-dead bodies indefinitely alive, the human body easily adapted to all forms of mechanical enhancement. What could emerge from the hypothetical patent on death and the ensuing sociotechnological shifts is a new phase of human-machine interface where the death of death becomes the condition of possibility for a human body that can die but may not remain dead.

But what would become of a society or civilization with radically reduced numbers of dead bodies? Any group of people that persists in unabated living would present substantial and stark economic problems for the larger population. Governmental care for the population would have to be entirely rethought, as death would no longer be one way to keep the nation's population under control. Per Foucault's thoughts on biopolitics, authorities' ability to control human life is one that requires the possibility of death, even if that death is ignored. What possible control can a sovereign yield if a population cannot easily die? Death, in theory, would no longer be a threat, but rather a kind of inconvenience for an already born population ultimately beyond the control of that sovereign. The politics of childbirth and population control would as well take on significantly different cultural stakes. Unless, of course, sovereign authorities are the ones responsible for managing which individuals can access prolonged life extension. This potentially new kind of battle over who or whom controls the end of life would surely make previous, similar fights look simple.

What would most likely happen with a technology stopping death, using the United States as an example, is that limited mortality would become an even more pervasive tool against the economically disadvantaged and technologically illiterate populations. Death would become a phenomenon largely affecting the poor, not unlike curable infectious diseases today, and would simultaneously privilege the affluent populations that could afford the technology. And while a labor force of lower socioeconomic groups would still be needed, the resources to keep those laboring bodies alive would become ever more scarce. These potential economic problems are reason enough not to pursue a patent that eliminates death.

The larger, more fundamental ethical problem created by any individual or sovereign authority attempting total control of human mortality is this: death represents a choice about life. And as human mortality leading to the limit of death becomes more and more technologized, that moment of choice, as Raymond Williams suggests, becomes itself more distributed and potentially uncontrollable.[17] Human beings would not be liberating themselves from dying as much as death, and the dead body would become separated by greater and greater degrees from the human. Foucault's articulation of the irony produced by various regimes controlling human sexuality that have us "believe that our 'liberation' is in the balance" also extends to the politics of death.[18] The biopolitical controls that are so apparent at every level of sovereign power today will soon merge with the powers of both necropolitics and thanatopolitics. In a world without death only life will exist, and living forever is surely no liberation. Indeed, it is the face of Man drawn in the sand washing out to sea.[19]

12/03/2018

Watching My Sister Die—Today Is My Birthday
Today is my birthday little sister.
the first one since you died
And we were together last year.
When I began to realize how much pain you
were in
and how bad the pain was becoming.
I look back now and think about
how much more I might have done
but didn't.
So I'm left now wondering
where your life might have gone
had I insisted sooner
that you knew
you were dying
But not on that trip
when all we did was laugh
and talk and take a photo
one of the last ones we took together
so today is my birthday little sister
and I don't much feel like celebrating
mostly because I'll live now   every year
without you telling me I'm getting old.

Coda: Planning for Death

Planning Julie's funeral. That was the other thing I wanted to do in July 2018 when I told Julie that she was dying. I even brought along official Centre for Death and Society (CDAS) funeral planning worksheets to help facilitate the conversation, death professional that I am. At one point, after talking about dying, I asked her about a funeral and if she had any specific wishes. She dryly asked me if I ever stopped working and then said no, she did not have any specific requests.

Julie explained that she would be dead and didn't really care because the funeral was for us. I made some quick notes and said okay. Then Julie said that she did have one request. She told me to make sure that no one turned her into some kind of hero during the funeral. She wanted people to remember her as a humble person, and what she was like in real life.

My sister truly resisted the epic hero narrative dropped on many people with cancer; she never felt heroic and would gladly choose the banality of everyday life over any glioblastoma multiforme–infused decline. Watching her transformed by both the cancer and its treatment made me tell my parents that I never wanted the devastating deterioration Julie experienced.

If ever diagnosed with brain cancer (however unlikely) or any other form of cancer, I would seriously question whether or not to pursue aggressive treatment. My parents feel the same way. I am also aware that cancers differ widely from each other; so universally refusing all possible treatments does not make total sense. I say this in the event the Global Union of Concerned Oncologists decides to contact me.

But brain cancer is different. It just relentlessly destroys the person. One of the classes I teach at the University of Bath is a final-year option called The Sociology of Death, and every year when we discuss definitions of death and dying, I ask my students, Where is the person located—in the brain or in the heart? I thought about that question when I discussed funeral planning with my sister since "Julie, the person" we all knew was disappearing before our eyes. The funeral was not important for her, not really. It was important for all of us to remember her in the memories we made before cancer.

Then Julie said something else, and I will never forget this end-of-life moment with my sister. I cherish it. She told me that if anyone at the funeral did transform her into a hero that she would be standing at the back of the room watching us and be really pissed off. Then she started swearing, and by swearing I mean *really* swearing. My mental notes during that exchange clocked a laundry list of choice profanity that in the moment both startled me and made me smile. Julie was definitely still inside that brain. She understood what I asked about. She got it and answered honestly. At both services in Italy and Wisconsin I told everyone Julie's one request: No hero talk.

Those unanswered CDAS funeral-planning sheets still sit in my journal as a reminder of things both accomplished and left undone before my sister died. CDAS developed the planning sheets over the years for use at academic conferences, at public engagement events, and in the classroom. I never expected to discuss those questions with my sister at that point in our lives, but that was also my own embarrassing death blind spot—any of us can die at any time. I grew up knowing this and regularly reminded people over the years that human mortality can quickly slip away. I earned a PhD saying these things. Talking about death and dying is literally my job. Julie and I certainly discussed and signed our parents' end-of-life decision-making and power of attorney for health care paperwork in June 2014 when they brought us the documents, but we excluded ourselves from that conversation. We also discussed our parents' funeral planning, which is an ever-evolving scenario but I think under control. I need to double-check. My Dad knows a guy.

So given everything that has happened to my own family, here are those Centre for Death and Society planning sheets, edited to include questions and considerations I know many people will confront when someone they know dies. Please add your own questions and concerns on anything you think is important. I decided to include the CDAS planning sheets for many reasons, but mostly because everyone should really think about all the questions they ask. Trust me when I say that most families do not discuss these points in advance. Funeral directors, hospice workers, bioethicists, and specialist palliative care doctors are often the worst offenders, which only makes a certain kind of ironic sense.

I also believe in paper-based funeral planning's technological simplicity. Many companies now offer online services to manage end-of-life decisions and funeral planning, but I am skeptical that most of those companies will exist in ten or twenty years. The online environment in which they operate changes too quickly. Paper forms will last as long as they are safely maintained, for example, in a folder, ideally also scanned onto a device that then also needs maintaining over time (see how complicated this all becomes), but here is the key part: your next of kin must know where to find this information. At a minimum discuss your responses with the people responsible for making these wishes known. It is also important to have broader family discussions about death and dying before someone is incapacitated, on life support, and medical care decisions need to happen. That all occurs before the funeral service disagreements can even begin. If anyone asks why you are asking all these death and dying questions, tell them the Overlord of Death said to do it.

I also recommend writing answers in pencil, since people often change their minds over time. So, for example, it is not uncommon to start thinking that some resuscitation might actually be okay as a person gets older and maintains good health. Being sixty looks a lot different to a twenty-year-old than a fifty-five-year-old. But let me repeat: make sure another person knows what you are thinking and definitely seek legal counsel if you think you need something more binding than the closing pages of my book.

END OF LIFE and FUNERAL PLANNING

Name:

Date:

Signature:

Indicate your funeral and end-of-life planning wishes by completing the following sections. Say as little or as much as you want about your choices.

- I want life support used (yes/no/maybe) and for what length of time? Under what conditions?

_____

_____

_____

_____

_____

_____

- Under what conditions should I be removed from life support? I nominate a specific person to make that decision should a decision need to be made (yes/no). Who?

_____

_____

_____

_____

_____

_____

- I leave all my funeral arrangement choices to this person or people:

_____

_____

_____

_____

_____

_____

- I (have/have not already) planned my funeral, and those arrangements are located at:

  _____

  _____

  _____

  _____

  _____

  _____

- I request that people do not do any of the following things at my funeral or memorial service:

  _____

  _____

  _____

  _____

  _____

  _____

- The funeral director, cemetery, crematorium/cremation society, or body disposal facility I prefer is:

  _____

  _____

  _____

  _____

  _____

  _____

- The price range I would like spent on my funeral is:

  _____

  _____

  _____

  _____

  _____

  _____

- I want my next of kin to choose my body's disposal method in the event I become incapacitated (yes/no/don't care):

  _____
  _____
  _____
  _____
  _____
  _____

- I would like my body (embalmed/not embalmed/kept on a cold pad/don't care):

  _____
  _____
  _____
  _____
  _____
  _____

- I would like my body (laid out at home/taken to a funeral home's visitation chapel/other/don't care) and for how long?

  _____
  _____
  _____
  _____
  _____
  _____

- I would like my body to be dressed in (formal clothes/specific outfit/don't care) for my funeral:

  _____
  _____
  _____
  _____
  _____
  _____

- I would like my body to be viewed by (my next of kin only/
  extended family/friends/work colleagues/anybody who
  wants to say goodbye):

  _____

  _____

  _____

  _____

  _____

  _____

- I would like my body to be transported to (a house of
  worship/secular remembrance space/crematorium/cemetery)
  in (a hearse/my own vehicle/other/don't care):

  _____

  _____

  _____

  _____

  _____

  _____

- If I am buried (or even if I'm not), I would like a gravestone
  or marker that says:

  _____

  _____

  _____

  _____

  _____

  _____

- If I am cremated, I want the following done with my ashes:

  _____

  _____

  _____

  _____

  _____

  _____

- I (do/do not) want flowers sent by (close family/friends/ anyone) to my funeral:

  _____

  _____

  _____

  _____

  _____

  _____

- I (do/do not) want monetary donations made to these orga-nizations in my memory:

  _____

  _____

  _____

  _____

  _____

  _____

- I would like a funeral or memorial service to be led by (religious minister—state which religion/celebrant/specific friend/specific family member/other):

  _____

  _____

  _____

  _____

  _____

  _____

- During the memorial service I would like the follow-ing music played (none/hymns/classical/music chosen by funeral director/special piece of music chosen by me/family choice/don't care):

  _____

  _____

  _____

_____

_____

_____

• I would like my organs, bones, and tissues donated (yes/
  no). Which specific ones?

_____

_____

_____

_____

_____

_____

• I have nominated someone to manage my online accounts
  and passwords (yes/no). Who? All online accounts or only
  certain ones? Does that person have the necessary logins
  and passwords?

_____

_____

_____

_____

_____

_____

• I want my funeral or memorial service to include the following things I have not already discussed in the other
  sections:

_____

_____

_____

_____

_____

_____

- The last thing I would like to say is:

_____
_____
_____
_____
_____
_____
_____
_____
_____
_____
_____

5/29/2019

Watching My Sister Die—The Last Page
The last page, little sister.
I've reached the last page for the book
you will never read
so I sat down on the couch in the house
the house you never saw
to say I've spent every day this year
thinking about your final year
About the   pain
and holding your hand
Watching Mom and Dad put flowers on your grave
Putting a stone next to the flowers every time I visit.
And you'll never read any of this little sister
I keep realizing this over and over
Every day as I ride my bike past Prometheus
wishing
That I didn't know death so well.
But I do.   I do.
So I watched you die little sister
saw you consumed by an unstoppable void.
Devouring you from the inside out
Even though the Doctors tried to save you like fire
from the Gods.
The last page, little sister. I am writing
    the last page.

Notes

Introduction

1. In January 2009, The President's Council on Bioethics released the following report: *Controversies in the Determination of Death: A White Paper by the President's Council on Bioethics*. The report largely supports the findings of the 1981 presidential commission. To view the report, see http://bioethics.georgetown.edu/pcbe/reports/death/index.html.

2. United States, President's Commission for the Study of Ethical Problems in Medicine and Biomedical Behavioral Research, *Defining Death: A Report on the Medical, Legal and Ethical Issues in the Determination of Death* (Washington, DC: Government Printing Office, 1981), 3.

3. President's Commission for the Study of Ethical Problems in Medicine and Biomedical Behavioral Research, *Defining Death*, 3.

4. The technologies of the human corpse concept is partially built around Michel Foucault's discussion of "human technologies." I discuss both the technologies of the human corpse and Foucault's human technologies at length in chap. 2. For Foucault's discussion of human technologies, see Michel Foucault, *Ethics, Subjectivity, and Truth*, ed. Paul Rabinow, trans. Robert Hurley (New York: New Press, 1997), 224–225.

5. Giorgio Agamben, *Homo Sacer*, trans. Daniel Heller-Roazen (Stanford: Stanford University Press, 1998), 164.

6. The *New York Times* printed a lengthy article in August 2005 about the issues confronting families when hospice or end-of-life care becomes necessary. The following quote is from the article: "As J. Donald Schumacher, president of the National Hospice and Palliative Care Organization, said last April to the Senate Committee on Health, Education, Labor and Pensions, 'Americans are more likely to talk to their children about safe sex and drugs than to their terminally ill parents about choices in care as they near life's final stages.'" To read the full article, see Robin Marantz Henig, "Will We Ever Arrive at the Good Death," *New York Times Magazine,* August 7, 2005, http://nytimes.com/2005/08/07/magazine/07DYINGL.html.

7. President's Commission for the Study of Ethical Problems in Medicine and Biomedical Behavioral Research, *Defining Death,* 3.

8. Alta Charo, "Dusk, Dawn, and Defining Death: Legal Classifications and Biological Categories," in *The Definition of Death: Contemporary Controversies,* ed. Stuart J. Youngner, Robert M. Arnold, and Renie Schapiro (Baltimore: Johns Hopkins University Press, 1999), 277.

9. Charo, "Dusk, Dawn, and Defining Death," 288.

10. University of Minnesota Center for Bioethics, *Determination of Death: Reading Packet on the Determination of Death* (Minneapolis: University of Minnesota, 1997), 4.

11. David Ewing Duncan wrote a succinct op-ed for the *New York Times* on the topic of life extension and asked a key question: How long do you want to live? See David Ewing Duncan, "How Long Do You Want to Live?" *New York Times,* August 25, 2012, https://nytimes.com/2012/08/26/sunday-review/how-long-do-you-want-to-live.html.

12. Agamben, *Homo Sacer,* 161.

13. T. Scott Gilligan and Thomas F. H. Stueve, *Mortuary Law,* 9th ed. (Cincinnati: Cincinnati Foundation for Mortuary Education, 1995), 6.

14. The key texts here are Jay Ruby, *Secure the Shadow* (Cambridge: MIT Press, 1995); Robert W. Habenstein and William M. Lamers, *The History of American Funeral Directing,* 4th ed. (Milwaukee: National Funeral

Directors Association of the United States, 1996); and Robert G. Mayer, *Embalming: History, Theory and Practice,* 3rd ed. (New York: McGraw-Hill, 2000).

15. Catherine Waldby, *AIDS and the Body Politic* (New York: Routledge, 1996).

16. Listen to the Gunther von Hagens radio story, "Cadaver Exhibits Are Part Science, Part Sideshow," on *National Public Radio,* August 10, 2006, at http://www.npr.org/templates/story/story.php?storyId=5553329.

17. In April 2012, Gunther von Hagens did open an exhibition of plastinated animal bodies at the Natural History Museum, London. That exhibition, *Animal Inside Out,* ran until September 2012. In 2018 he opened a tourist themed *Body Worlds* exhibition in Piccadilly Circus, also in London.

18. Annie Cheney's *Body Brokers: Inside America's Underground Trade in Human Remains* (New York: Broadway Books, 2006) is an important investigative text on contemporary American body brokers and offers important insights into the financial greed that motivates the industry. Michael Sappol's *A Traffic of Dead Bodies: Anatomy and Embodied Social Identity in Nineteenth-Century America* (Princeton: Princeton University Press, 2002) provides a more complex analysis of body brokering as a historical phenomenon.

19. Giorgio Agamben, *Remnants of Auschwitz,* trans. Daniel Heller-Roazen (New York: Zone Books, 1999), 83–84.

20. The concept of necropolitics has emerged in other writings, but those authors confuse death and the dead body. Necropolitics is a political situation involving the dead body and not necessarily death. See both *Polygraph 18: Biopolitics, Narrative, Temporality,* issue eds. Rodger Frey and Alexander Ruch (2006), and Achille Mbembe's "Necropolitics," trans. Libby Meintjes, *Public Culture* 15, no. 1 (2003): 11–40.

21. Raymond Williams, *The Politics of Modernism: Against the New Conformists* (New York: Verso, 1989), 134.

Chapter 1

1. *Casket*, December 1902, 30–31. Copies of *The Casket* are in the archives of the National Funeral Directors Association in Milwaukee, Wisconsin.

2. *Casket*, 30.

3. Over the last fifteen years books such as Mary Roach's *Stiff: The Curious Lives of Human Cadavers* (New York: W. W. Norton, 2003), Gary Laderman's *Rest in Peace: A Cultural History of Death and the Funeral Home in Twentieth-Century America* (New York: Oxford University Press, 2003), and Norman Cantor's *After We Die: The Life and Times of The Human Cadaver* (Washington, DC: Georgetown University Press, 2010) have been published, and these texts certainly discuss the human corpse. It is interesting to note however that discussions of death rarely tackle *how* the changes made to the human corpse have also affected fields like thanatology.

4. The President's Commission for the Study of Ethical Problems in Medicine and Biomedical Behavioral Research and its report *Defining Death: A Report on the Medical, Legal, and Ethical Issues in the Determination of Death* (1981), discussed in the introduction, is an example of a text combing through the countless definitions of death in contemporary America.

5. While many theorists, including Jean-François Lyotard and Fredric Jameson, have explored the postmodern subject, it is the *postmortem* subject that is this project's central body. The postmortem can and always will exist without the concept of the postmodern, but without death the question of what can ever be beyond the modern is more complex. Here I am thinking specifically of Jean-François Lyotard's *The Postmodern Condition: A Report on Knowledge*, trans. Geoff Bennington and Brian Massumi (Minneapolis: University of Minnesota Press, 1984), and of Fredric Jameson's *Postmodernism* (Durham: Duke University Press, 1991).

6. I am not suggesting that the human corpse has had a seamless movement over the last 150 years, but rather that as the technologies affecting

dead bodies have changed, so too have the postmortem conditions of death. We currently live in a "digital death" technology moment, but how long this moment lasts or is something people even remember in the years to come remains to be seen. These digital technology innovations reflect the current set of communication tools humans can use to discuss death, dying, and dead bodies. As with postmortem photography, the use of these digital tools will be far more compelling in one hundred years' time when what happened online is rediscovered as something new.

7. According to T. Scott Gilligan and Thomas F. H. Stueve in their book *Mortuary Law*, 9th ed. (Cincinnati: Cincinnati Foundation for Mortuary Education, 1995), 5, "The term 'dead body' means specifically the body of a human being deprived of life but not yet entirely disintegrated. The term 'corpse' is synonymous with the term 'dead body.' A body to be legally a dead body or corpse must meet three conditions: it must be the body of a human being, without life, and not entirely disintegrated." A human corpse that has decomposed to the point of organic remnants and/or become skeletal remains is no longer recognized as a dead body. Gilligan and Stueve give the following example from US case law: "In *State v. Glass*, 27 O. App. 2d 214, 273 N.E. 2d 893, a real estate developer who had ordered bulldozers to level land upon which an old cemetery was located was charged with a violation of Ohio's 'Grave Robbery' statute. The site that was developed contained the graves of three persons buried about 120 years earlier. In reversing the conviction of the developer, the Appeals Court stated as follows: 'A cadaver is not an everlasting thing. After undergoing an undefined degree of decomposition, it ceases to be a dead body in the eyes of the law'" *Mortuary Law*, 9th ed., 5.

8. While I do not make explicit use of either Walter Benjamin's "The Work of Art in the Age of Mechanical Reproduction" or "A Short History of Photography," both texts have been important in my thinking about death photography. *Camera Lucida* (1980) by Roland Barthes is also an interesting reference for the question of death and photography.

9. Tom Gunning, "Phantom Images and Modern Manifestations," in *Fugitive Images: From Photography to Video*, ed. Patrice Petro (Bloomington: Indiana University Press, 1995), 66.

10. A few substantial books have been published on the subject of death (also called postmortem) photography. The following works are important for both their visual documentation of the images and the historical information associated with the practice of photographing the dead. These books include Jay Ruby's *Secure the Shadow* (Cambridge: MIT Press, 1995), James Van Der Zee's *The Harlem Book of the Dead* (Dobbs Ferry, NY: Morgan & Morgan, 1978), Michael Lesy's *Wisconsin Death Trip* (New York: Pantheon Books, 1973), and two books by Stanley Burns, *Sleeping Beauties: Memorial Photography in America* (New York: Burns Archive Press, 1990) and *Sleeping Beauty II: Grief, Bereavement and the Family in Memorial Photography* (New York: Burns Archive Press, 2002).

11. Gunning, "Phantom Images," 48.

12. Gunning, "Phantom Images," 64. Gunning does an excellent job of detailing the growth of "spirit photography" during the nineteenth century. Spirit photographs were, as Gunning states, "produced as mourning images during the nineteenth century" (66). To achieve this effect, photographers would superimpose one image over another to create the appearance of a ghost-like specter.

13. Stanley Burns talks briefly about the process of developing these images as well as the role of P. T. Barnum in debunking them in *Sleeping Beauty II: Grief, Bereavement and the Family in Memorial Photography* (2002).

14. Ruby, *Secure the Shadow*. Ruby's book is an interesting synthesis and documentary source for all kinds of postmortem photography, including both humans and pets.

15. Ruby, *Secure the Shadow*, 52.

16. See Richard Leppert, *Art and the Committed Eye: The Cultural Functions of Imagery* (Boulder, CO: Westview Press, 1996).

17. Ruby, *Secure the Shadow,* 50. It's worth noting that an excellent source of African American postmortem photography from the early-to-mid-twentieth century is James Van Der Zee's *Harlem Book of the Dead,* noted above.

18. Gunning, "Phantom Images," 68.

19. An excellent essay on the prephotographic and pre-embalming spectacle of the human corpse is Vanessa Schwartz's "Cinematic Spectatorship before the Apparatus: The Public Taste for Reality in *Fin-de-Siècle* Paris," in *Cinema and the Invention of Modern Life,* ed. Leo Charney and Vanessa R. Schwartz (Berkeley: University of California Press, 1995), 297–319; pp. 298–304 focus on the general public's fascination with looking at unclaimed corpses in the Paris Morgue.

20. Ruby, *Secure the Shadow,* 53.

21. Ruby, *Secure the Shadow,* 59.

22. Ruby, *Secure the Shadow,* 59.

23. Robert W. Habenstein and William M. Lamers, *The History of American Funeral Directing,* 4th ed. (Milwaukee: National Funeral Directors Association of the United States, 1996), 212.

24. Habenstein and Lamers, *The History of American Funeral Directing,* 212.

25. Habenstein and Lamers, *The History of American Funeral Directing,* 219. Because of both price gouging and poorly done battlefield embalming, the Union Army received a War Department General Order in March 1865, entitled the "Order Concerning Embalmers," which mandated using only licensed, competent embalmers for war dead. The Civil War then ended a month later, but the mandate was the first time any formal governmental licensing took effect. Habenstein and Lamers discuss this entire history in their book on pp. 207–219.

26. Habenstein and Lamers, *The History of American Funeral Directing,* 217.

27. Carl Lewis Barnes, *The Art and Science of Embalming* (Chicago: Trade Periodical, 1896), 143.

28. Barnes, *The Art and Science of Embalming,* 183.

29. Habenstein and Lamers talk about public resistance to embalming during both the pre– and post–Civil War years because of religious concerns. As they explain, "Public resistance to the mutilation of the remains, moreover, had the sanction of the Christian tradition that the body is the temple of God, and that the remains are always sacred and must in every case be treated with reverence." See *The History of American Funeral Directing,* 218. Because of the public concerns, funeral directors such as Barnes often compared their profession to doing "God's work." Interestingly, in 1910 at the 7th Annual Conference of the Embalmers' Examining Boards of North America, a Rev. John H. Nawn made a formal address regarding the theological importance of the embalmers' work entitled "The Sacredness of Our Profession." See the "Proceedings of the 7th Annual Conference of Embalmers' Examining Boards of North America," pp. 6–8.

30. James Farrell, *Inventing the American Way of Death, 1830–1912* (Philadelphia: Temple University Press, 1980), 160–161.

31. Farrell, *Inventing the American Way of Death,* 158.

32. Habenstein and Lamers, *The History of American Funeral Directing,* 303.

33. Habenstein and Lamers, *The History of American Funeral Directing,* 303.

34. "Proceedings of the 4th Annual Joint Conference of Embalmers' Examining Boards of North America," Norfolk, VA (1907), 9–10.

35. "Proceedings of the 4th Annual Joint Conference of Embalmers' Examining Boards of North America," Norfolk, VA (1907), 9–10.

36. Late nineteenth- and early-twentieth-century funeral industry periodicals such as *The Sunnyside, The Casket,* and *The Embalmer's Monthly* are littered with stories about "fluid Men."

37. Habenstein and Lamers, *The History of American Funeral Directing,* 219.

38. To offer a sense of how long and drawn out the process of developing the regulations for transcontinental shipment of dead bodies was, the National Association of General Baggage Agents first drew up suggested rules for shipping bodies in 1888. See Habenstein and Lamers, *The History of American Funeral Directing,* 319.

39. The information about this meeting and the rules agreed upon by the different groups were reprinted in the 1906 "Proceedings of the 3rd Annual Meeting of Association of State and Provincial Boards of Health and Embalmers' Examining Boards of North America," 46–48.

40. "Proceedings of the 3rd Annual Meeting of Association of State and Provincial Boards of Health and Embalmers' Examining Boards of North America," 46–47.

41. Also see Michel Foucault's discussion of "cadaveric time" in Michel Foucault, *Birth of the Clinic,* trans. Alan Sheridan (New York: Vintage, 1973), 141.

42. Wolfgang Schivelbusch, *The Railway Journey; the Industrialization of Time and Space in the 19th Century* (Berkeley: University of California Press, 1977), 33.

43. Schivelbusch, *The Railway Journey,* 35.

44. Habenstein and Lamers, *The History of American Funeral Directing,* 320.

45. Carl Lewis Barnes is one of the more colorful figures in the history of American embalming. Habenstein and Lamers refer to him as "that showman, Dr. Carl Lewis Barnes" (p. 328) in *The History of American Funeral Directing.* In Robert Mayer's *Embalming: History, Theory and Practice,* 3rd ed. (New York: McGraw-Hill, 2000), 475, he gives the following biographical sketch of Barnes:

> Born into a family that operated an undertaking establishment in Connellsville, Pennsylvania, Barnes (1872–1927) studied medicine in Indiana, opened an embalming school there and moved it to Chicago. He manufactured

embalming chemicals, wrote many books and articles on the subject, and had the largest chain of fixed-location schools in history in New York, Chicago, Boston, Minneapolis and Dallas. While serving overseas as a medical colonel in the US Army in World War I, his businesses failed. He never reopened the schools, continuing the practice of medicine until his death.

46. The Bisga Fluid advertisements ran in a series of embalming and funeral trade journals, namely, *The Casket* and *The Sunnyside* from 1902–1903, and are available in copies of those journals at the National Funeral Directors Association headquarters in Milwaukee, Wisconsin. These publications were not printed for reading by the general public, so it is doubtful that many people other than funeral directors or embalmers saw the ads. What is unknown about the Bisga Man is where exactly Carl Lewis Barnes obtained the body. Barnes makes a brief reference to the man's corpse featured in the advertisements during a lecture in 1905 to the Connecticut Funeral Directors Association. According to a typewritten account of the lecture in the National Funeral Directors Association archive, Barnes says the following about embalming "difficult cases" (meaning bodies ravaged by disease or in a badly decomposed state): "The best case I ever treated was a failure at the start. The body was that of a young man about thirty years of age" (15). Barnes alludes to the idea that the young man was killed by tuberculosis and that he had great difficulty injecting fluid into the dead body. After successfully embalming the body, Barnes explains, "I used a picture of this subject in advertising three months after death. It [the body] would have gone to pieces in three days if it had not been so treated" (16). Everything else about the Bisga Man is somewhat of a mystery.

47. This statement is taken from a December 1902 version of the Bisga Man advertisement in *The Casket*. The advertisement is in the archives of the National Funeral Directors Association.

48. Carl Lewis Barnes wanted funeral directors to realize how good dead bodies could look using *his* products. Barnes ran another advertisement in 1906 for Bisga Embalming Fluid in which the selling point was the following line: "WAIT UNTIL THE Mercury Reaches 90 Degrees Fahrenheit, Then you will wish you had some BISGA." This advertisement is in *The Sunnyside*.

Chapter 2

1. See United States, President's Commission for the Study of Ethical Problems in Medicine and Biomedical Behavioral Research, *Defining Death: A Report on the Medical, Legal and Ethical Issues in the Determination of Death* (Washington, DC: Government Printing Office, 1981).

2. See United States, President's Council on Bioethics, *Controversies in the Determination of Death: A White Paper by the President's Council on Bioethics* (Washington, DC: Government Printing Office, 2009), http:// bioethics.georgetown.edu/pcbe/reports/death/index.html.

3. Lyn Lofland, *The Craft of Dying: The Modern Face of Death,* 40th anniversary ed. (Cambridge: The MIT Press, 2019), 2.

4. See Peter J. Donaldson, "Denying Death: A Note Regarding Ambiguities in the Current Discussion," *Omega,* November 1972, 285–290. Richard G. Dumont and Dennis C. Foss, *The American View of Death: Acceptance or Denial?* (Cambridge: Schenkman, 1972).

5. Lofland, *The Craft of Dying,* 72–73.

6. Lofland, *The Craft of Dying,* 86.

Chapter 3

1. The National Funeral Directors Association, based in Milwaukee, Wisconsin, was formed in 1880. The NFDA website (http://www.nfda .org) gives the following summary of the organization's history: "On January 14, 1880, a group of 26 Michigan undertakers held the first meeting of what would be established in 1882 as the National Funeral Directors Association (NFDA). Today, NFDA is the oldest and largest national funeral service organization in the world."

2. National Funeral Directors Association Memorandum, *AIDS Precautions for Funeral Service Personnel and Others,* June 1985, 1. The impetus for this memo also came from Dr. John H. Richardson at the Centers for Disease Control in Atlanta, Georgia, when he asked the NFDA to "send

him material funeral service has available offering advice on the protection of funeral service personnel."

3. National Funeral Directors Association, *AIDS Precautions*, 1.

4. The eight "embalming precautions" are as follows: "(1) Embalmers should wear double rubber gloves and a disposable apron, plus a mask making sure that their hair is covered. (2) Instruments should be washed in 1:10 Clorox solution. (3) Aprons and gloves used during embalming should be placed in a plastic bag and incinerated as soon as possible. (4) Use goggles or a pair of glasses. (5) Use shoe covers and dispose of along with the gloves and apron. (6) The table and floor should be washed with the Clorox solution. (7) Rags and towels used during the preparation should be destroyed. (8) If the body is being viewed, the family should avoid having physical contact with it."

5. In order to determine whether the eight embalming precautions remained in use, I asked former University of Minnesota Program of Mortuary Science Embalming Instructor Jody LaCourt about each one. In an e-mail message, LaCourt explained:

> You're right, embalming precautions have not changed very much over the last 20 years. However, the OSHA [Occupational Safety and Health Administration] training, which includes proper PPE [personal protective equipment,] that my students and the Minnesota funeral directors receive goes into greater detail than the eight precautions you included in your email. Along with an apron and gloves we are also required to wear a disposable gown, which covers all exposed skin, as well as goggles that fit over our glasses. This may be a minor detail, but I just thought I would mention it. You are also correct about the fact that families are no longer being told to avoid physical contact with the body. I can't imagine telling the family that they must not touch their loved one's hand or forehead. How upsetting that must have been for the families who were told such a statement. I cannot speak for other funeral directors, but the one change that I have noticed in the last 10–15 years has nothing to do with the eight embalming precautions when embalming a body that has died from HIV/AIDS, but it is the attitudes of funeral directors that have changed. (Personal e-mail to author [September 2009])

6. In 1992, the Infectious/Contagious Disease Committee of the Funeral Directors Services Association of Greater Chicago commissioned a study that asked, How long does HIV survive in a dead human body? The 1992

study (published in the January 1993 issue of *The Director*) is discussed at the end of this chapter, but the opening lines of the study merit note here. The report's authors explain, "The risk of contamination from infected blood or bodily fluids is an extremely important consideration for those who work with deceased AIDS patients. ... Rumors abound: the virus dies immediately, within 24 hours, after seven days, or is killed by refrigeration. Funeral directors desperately need factual evidence on which to base their actions and set their minds at ease, but none has been forthcoming. Until now." Funeral Directors Services Association of Greater Chicago, "AIDS Update," *The Director*, January 1993, 56–57.

7. The Program of Mortuary Science at the University of Minnesota is one of the oldest programs of its kind in America. On November 1, 2008, the Program of Mortuary Science hosted a Centennial Celebration. For more information on the University of Minnesota Mortuary Science degree, go to https://www.mortuaryscience.umn.edu.

8. Jody LaCourt, personal e-mail to author (September 2009).

9. The term "direct disposition" means immediately burying, cremating, or legally disposing of the dead body without any form of embalming or public ceremony.

10. National Funeral Directors Association Board of Governors, "Acquired Immune Deficiency Syndrome Policy of the National Funeral Directors Association of the United States, Inc.," *The Director*, October 1985, 19.

11. Catherine Waldby, *AIDS and the Body Politic* (New York: Routledge, 1996), 1.

12. Robert G. Mayer, *Embalming: History, Theory and Practice,* 3rd ed. (New York: McGraw-Hill, 2000), xiv.

13. The National Funeral Directors Association, *The Director*, June 1986, 46.

14. Clarence G. Strub and L. G. "Darko" Frederick, *The Principles and Practice of Embalming,* 5th ed. (Dallas: Professional Training Schools, 1989), 3.

15. Paul Rabinow, *Essays on the Anthropology of Reason* (Princeton: Princeton University Press, 1996), 36.

16. Robert G. Mayer, "Offering a Traditional Funeral to All Families," *The Director*, September 1987, 28–30.

17. Mayer, "Offering a Traditional Funeral to All Families," 28–30.

18. Throughout the articles and industry literature I compiled for this chapter, any reference to an unnamed funeral director who would not embalm an HIV/AIDS corpse is usually countered by mentioning another unnamed funeral director who would. I mention this point because it is clear from the articles I collected that the funeral industry was really grappling with how to handle the situations caused by homophobia, fear of AIDS, and overall institutional changes caused by the virus during the 1980s. This was certainly the case in the already discussed 1986 Q and A about New York State's embalming laws.

19. Both Mayer in *Embalming: History, Theory and Practice*, 3rd ed., and Strub and Frederick in *The Principles and Practice of Embalming*, 5th ed., discuss the radiation danger in terms of both industrial accidents *and* the medical uses of radiation in cancer treatments. In Mayer, see pp. 412–414 (with special attention to his mention of *Thanatochemistry*), and in Strub and Frederick, see pp. 348–352. To give a further example of these radiation exposure concerns, Strub and Frederick began chap. 29 of their book, which was succinctly titled RADIATION CASES and published in 1989, this way:

> The care and preparation of radioactive bodies have in recent years become a matter of major interest and importance for the embalmer. This is due primarily to disasters involving nuclear plants and the release of radiation. As nuclear technology grows, the possibility of more nuclear accidents increases. Not only does this factor impose a health hazard equal to that presented by the more serious infections, but it also creates the possibility that there will be a long and objectionable time interval between death and embalming. (348)

20. Mayer, *Embalming: History, Theory and Practice*, 413. The radiation safety officer needs to sign off on a special form that accompanies a radioactive body saying the corpse is safe to handle in the funeral home.

Funeral directors can as well become certified to handle radiation accident victims in the event of a nuclear disaster.

21. In 1992 the National Commission on Acquired Immune Deficiency Syndrome published "America Living with AIDS," and the commission's recommendations for the American funeral industry appeared in *The Director*, January 1992, 18–22. I mention the report because it helped clarify that while precautions were needed when embalming HIV/AIDS corpses, the dangers represented by AIDS were not nearly as severe as many funeral directors believed.

22. Michel Foucault, *Ethics, Subjectivity, and Truth*, ed. Paul Rabinow, trans. Robert Hurley (New York: New Press, 1997), 224–225.

23. Foucault, *Ethics, Subjectivity, and Truth*, 225.

24. *Technologies of the self* is something Michel Foucault discusses in various publications. I am working with the "Technologies of the Self" seminar Foucault gave at the University of Vermont in 1982. A text based on that seminar can be found in Foucault's *Ethics, Subjectivity, and Truth*, 223–251. Another location in which "technologies of the self" or "cultivation of the self" (as it is sometimes translated) can be found is in Foucault's *The History of Sexuality, Volume III*, trans. Robert Hurley (New York: Vintage Books, 1986), 43–45.

25. See James H. Bedino, *AIDS: A Comprehensive Update for Embalmers*, Research and Education Department of The Champion Company, No. 616 (1993).

26. Both Mayer, in *Embalming: History, Theory and Practice*, 3rd ed., and Strub and Frederick, in *The Principles and Practice of Embalming*, 5th ed., talk about the importance of tagging dead bodies so the embalmer and/or funeral home employee knows what the cause of death was when it is not clear. Neither book gives a history of when the tagging began, but the use of COD (cause-of-death) tags was something that gained added importance when handling the HIV/AIDS corpse.

27. Mayer, "Offering a Traditional Funeral to All Families," 28–30.

28. The following information about "universal precautions" was taken from the OSHA website: "On December 6, 1991, the Occupational Safety and Health Administration (OSHA) promulgated the Occupational Exposure to Bloodborne Pathogens Standard. This standard is designed to protect approximately 5.6 million workers in the health care and related occupations from the risk of exposure to bloodborne pathogens, such as the Human Immunodeficiency Virus and the Hepatitis B Virus." For more information on universal precautions, see the www.osha.gov website.

29. Mayer, *Embalming: History, Theory and Practice*, 37.

30. Dalton Sanders, "Err on the Side of Caution," *The Director*, April 1997, 73–74.

31. Foucault, *Ethics, Subjectivity, and Truth*, 51.

32. On the figure of the *human monster*, see Michel Foucault, *Abnormal: Lectures at the Collège de France, 1974–1975*, trans. Graham Burchell (New York: Picador, 2003), 55–56.

33. Jerome F. Frederick, "AIDS—Identification and Preparation," *The Director*, July 1985, 8–11, 43–44.

34. Waldby, *AIDS and the Body Politic*, 140.

35. Waldby, *AIDS and the Body Politic*, 145.

36. Waldby, *AIDS and the Body Politic,* 146.

37. Michael Hearn, "Photographs and Memories," *The Director*, January 1992, 10–13, 57–58. For another similar description of an HIV/AIDS funeral where the contested meanings and control of the dead body took place, see chap. 1 of Simon Watney's *Policing Desire: Pornography, AIDS and the Media* (Minneapolis: University of Minnesota, 1996).

38. Michael Lensing, "Arrangement Conference for AIDS Related Deaths," *The Director*, December 1996, 6–8.

39. T. Scott Gilligan and Thomas F. H. Stueve, *Mortuary Law,* 9th ed. (Cincinnati: Cincinnati Foundation for Mortuary Education, 1995), 6.

40. The legal claim to a decedent's remains is one of the many issues involved with making same-sex marriage legal in the United States. In *Obergefell v. Hodges* (2015), the US Supreme Court ruled that same-sex marriage applied across the United States, and that ruling legally settled who could claim the decedent's remains. Without this recognition, same-sex partners were not legally next of kin. Jim Obergefell originally filed his legal case because he was prevented from being listed as "spouse" on his husband John Arthur's death certificate. Obergefell and Arthur were married in Maryland but lived in Ohio (where their marriage was not legally recognized). The backstory as to why and how the case made it to the US Supreme Court is often forgotten. See this short 2015 *New York Times* video documentary about the case: *How a Lover Story Triumphed in Court:* https://www.nytimes.com/video/us/100000003765330/how-a-love-story-triumphed-in-court.html.

41. Kath Weston, *Families We Choose: Lesbians, Gays, Kinship* (New York: Columbia University Press, 1991), 56.

42. See Simon Watney, *Imagine Hope: AIDS and Gay Identity* (London: Routledge, 2000), and his discussion of "political funerals."

43. Foucault, *The History of Sexuality, Volume I*, 135.

44. Waldby, *AIDS and the Body Politic*, 146.

45. Funeral Directors Services Association of Greater Chicago, "AIDS Update," *The Director*, January 1993, 56–57.

46. Public health concerns over newer kinds of "dangerous" corpses persist today. Both the Ebola corpse and the Creutzfeldt-Jakob disease corpse (the human variant of Mad Cow disease) represent their own specific kinds of infection and lethality. The HIV/AIDS corpse again becomes productive in these scenarios, since funeral directors and public health workers negotiate how to safely handle the human remains while also attempting to provide a funeral. The cultural politics, especially around Ebola and the history of colonialism, are extremely complicated and often demonstrate how quickly the politics of death emerge when trying to follow best practices that reduce further infections.

47. Jody LaCourt provided these comments on current practices:

Moving ahead ... I can proudly say that I no longer see funeral directors "freaking out" about embalming an HIV/AIDS body. They embalm these bodies just like they normally would embalm a body—they practice using universal precautions. I feel the main reason for this is simply the fact that funeral directors are more knowledgeable about HIV/AIDS. We used to be afraid of touching a human body that was HIV positive, and I feel those feelings stemmed from the unknown. We did not know much about HIV/AIDS or its transmission back then. We thought we knew that we were at an extremely high risk for contracting AIDS if we were to embalm a body with this disease. Fear of the unknown is common in human nature, and it's amazing what happens to fear when we become more knowledgeable. Once we have all the facts, fear amazingly subsides. (Jody LaCourt, personal e-mail to author September 2009)

Chapter 4

1. From Institute for Plastination, *Donating Your Body for Plastination*, 7th rev. ed., December 2004, 23.

2. Listen to the Gunther von Hagens radio story, "Cadaver Exhibits Are Part Science, Part Sideshow," on *National Public Radio*, August 10, 2006, at http://www.npr.org/templates/story/story.php?storyId=5553329.

3. See "What Is *Body Worlds*" in the FAQ section of the *Body Worlds* website for more information on visitor attendance: https://bodyworlds .com/about/faq/.

4. Institute for Plastination, *Donating Your Body for Plastination*, 8.

5. AFP, "Copulating Corpses Raise the Roof in Berlin," May 7, 2009, https://www.france24.com/en/20090507-exhibition-germany-doctor -death-copulating-corpses-raise-roof-berlin-hagens-anatomy. Also see Dave Itzkoff, "Cadaver Sex Exhibition in Germany Is Criticized," *New York Times,* May 7, 2009, https://www.nytimes.com/2009/05/08/arts/ design/08arts-CADAVERSEXEX_BRF.html. And see "Copulating Corpses Spark Outrage in Berlin Show," *Reuters*, May 6, 2009, https://www.reuters .com/article/us-finearts-corpses/copulating-corpses-spark-outrage-in -berlin-show-idUSTRE5455CI20090506.

6. Here is von Hagen's advertorial letter, making his appeal to the people of London. It was published on June 23, 2009, in the *London Evening Standard.*

Open Letter from Gunther von Hagens

Dear Reader,

Many of you know me as the anatomist who brought the post-mortal body into public consciousness. In recent weeks, politicians and pundits have questioned my latest work: the anatomical preservation of a man and a woman—two consenting, deceased donors—through my science of plastination, in a pose meant to highlight human reproduction.

Ironically, despite the criticism of the powerful, our visitors have supported the by now world-renowned "Sex Couple." Since the unveiling of this display in Berlin, we have received hundreds of messages from the UK urging us to present such a special plastinate in the BODY WORLDS exhibition currently on show at The O2 bubble.

As a public anatomist, I have strived throughout my career to listen to the people rather than to the arbiters of taste and propriety. I therefore call upon you, the Great British public, to tell me your thoughts on this plastinate. Would you like to see a similar plastinate come to London and permanently join the BODY WORLDS & The Mirror of Time exhibit at The O2?

Please join the debate and cast your vote.

The plastinate which is currently in Germany is ready to make its way to London. If you vote now, this unique display could be permanently in the BODY WORLDS exhibition at The O2 bubble in a matter of weeks.

Please cast your vote and help me make my decision.

Best Wishes,

Gunther

To read von Hagens's appeal online, see http://www.thisislondon .co.uk/standard-home/body-worlds-sex-couple-the-debate-6801712 .html.

7. In April 2012, Gunther von Hagens did open an exhibition of plastinated animal bodies at the Natural History Museum, London. That exhibition, *Animal Inside Out*, ran until September 2012. In 2018 he opened a tourist-themed *Body Worlds* exhibition in Piccadilly Circus, also in London.

8. See Tony Walter, "Body Worlds: Clinical Detachment and Anatomical Awe," *Sociology of Health & Illness* 26, no. 4 (2004): 464–488. One of the controversies associated with an early incarnation of *Body Worlds* involved the lack of female bodies on display. The absence of female bodies led some spectators to complain. Tony Walter explains:

> The one, major, exception to Body World's sense of common humanity is the exclusion of the female body. Apart from the swimmer, the only female whole-body plastinates display aspects of reproduction, implicitly defining the female as a reproductive machine. Many females write that they would like to learn about themselves as well as about men … Von Hagens['s] … defence for excluding female whole plastinates is that he does not want Body Worlds to be accused of enabling males to become voyeurs of women's bodies. (483)

9. Some of these institutions include the San Diego Natural History Museum; the Minnesota Science Museum; the Buffalo Museum of Science; the Great Lakes Science Center in Cleveland, Ohio; the California Science Center in Los Angeles; and the Milwaukee Public Museum.

10. Michael Sappol, *A Traffic of Dead Bodies: Anatomy and Embodied Social Identity in Nineteenth-Century America* (Princeton: Princeton University Press, 2004), 203.

11. Kate Connolly, "Fury at Exhibit of Corpses Having Sex," *Guardian*, May 6, 2009, https://www.theguardian.com/world/2009/may/06/german-artist-sex-death.

12. Connolly, "Fury at Exhibit of Corpses Having Sex," https://www.theguardian.com/world/2009/may/06/german-artist-sex-death.

13. AFP, "Copulating Corpses Raise the Roof in Berlin," https://www.france24.com/en/20090507-exhibition-germany-doctor-death-copulating-corpses-raise-roof-berlin-hagens-anatomy.

14. Reuters, "Copulating Corpses Spark Outrage in Berlin Show," https://www.reuters.com/article/us-finearts-corpses/copulating-corpses-spark-outrage-in-berlin-show-idUSTRE5455CI20090506.

15. The full title of Linnaeus's book is *System of Nature through the Three Kingdoms of Nature, According to Classes, Orders, Genera and Species, with [Generic] Characters, [Specific] Differences, Synonyms, Places.* See Giorgio Agamben, *The Open* (Stanford: Stanford University Press, 2004), 25–26.

16. Agamben, *The Open*, 25–26.

17. See Walter, "Body Worlds," 464–488.

18. Institute for Plastination, *Donating Your Body for Plastination*, 13–14.

19. Gunther von Hagens, *KÖRPERWELTEN Exhibition Guide*, 4th ed. (2001): 12.

20. Walter, "Body Worlds," 478.

21. Michel Foucault, *Birth of the Clinic*, trans. Alan Sheridan (New York: Vintage, 1973).

22. Foucault, *Birth of the Clinic*, 141.

23. Georges Canguilhem, *The Normal and the Pathological*, trans. Carolyn R. Fawcett (New York: Zone Books, 1989), 237–239.

24. Paolo Virno, *A Grammar of the Multitude*, trans. Isabella Bertoletti, James Cascaito, and Andrea Casson (New York: Semiotext[e], 2004), 81.

25. Von Hagens, *KÖRPERWELTEN Exhibition Guide*, 18.

26. Edward Tyson, one such early taxonomist, published the following treatise in 1699: *Orang-Outang, sive Homo Sylvestris: or, the Anatomy of a Pygmie Compared with that of a Monkey, an Ape, and a Man. To which is added, a Philosophical Chapter Concerning the Pygmies, the Cynocephali, the Satyrs, and Sphinges of the Ancients. Wherein it Will Appear that They are all Either Apes or Monkeys, and not Men, as formerly Pretended.* See Agamben, *The Open*, 25.

27. It also helps that through this process of reinvention Gunther von Hagens manages to accrue millions of dollars in ticket sales, an always important acknowledgement of his skills as a showman, à la P. T. Barnum.

28. Michel de Certeau, *Heterologies: Discourse on the Other*, trans. Brian Massumi (Minneapolis: University of Minnesota Press, 1986), 201–202.

29. Institute for Plastination *Body Donation Program* release forms.

30. Institute for Plastination *Body Donation Program*, 2.

31. The donation form does stipulate that:

> I understand that this is a legal document being signed by me (or at my direction by another) in accordance with the Uniform Anatomical Gift Act or similar laws. I can withdraw my consent for my body to be used for Plastination at any time without having to give a reason, but that I must do so in writing and signed by two witnesses. The IfP also has the right to issue a statement withdrawing its agreement to accept a body for Plastination. (2)

32. *Body Worlds* Press Release, "North American Donors to *Body Worlds* Anatomical Exhibitions Converge in Los Angeles to Meet Scientist, Gunther von Hagens & Discuss Their Post-Mortal Lives" (June, 10, 2008).

33. Institute for Plastination, *Donating Your Body for Plastination*, 5.

Chapter 5

1. Martha W. Anderson and Renie Schapiro, "From Donor to Recipient: The Pathway and Business of Donated Tissues," in *Transplanting Human Tissue: Ethics, Policy, and Practice*, ed. Stuart J. Youngner, Martha W. Anderson, and Renie Schapiro (Oxford: Oxford University Press, 2004), 13.

2. On December 1, 2011, a three-judge panel of the US Court of Appeals, Ninth Circuit, ruled that a donor of bone marrow could be legally paid for that biological material. The 1984 National Organ Transplant Act includes bone marrow as a biological material that cannot be sold, and such a transaction could be a felony. The case was being appealed by the US Justice Department. See *Flynn v. Holder*, 655 F. 3d 1048 (Ninth Cir., 2011). It is also important to point out that US organ laws have not prevented other countries from the selling of transplantable organs, namely, India, China, Pakistan, the Philippines, and Iran. Entire economies have developed as a result of these markets (which are sometimes legal and sometimes not) and produced a burgeoning medical tourism industry. In parts of India, for example, the street value of a kidney can be around $3,000 and that same kidney will be resold to a Westerner for $85,000 with none of the profit going to the donor. See Scott Carney, "Why a Kidney (Street Value $3,000) Sells for $85,000," *Wired*, May 8, 2007, https://www.wired.com/2007/05/india-transplants-prices. The work of

Nancy Scheper-Hughes at the University of California, Berkeley, has also made the ethical and medical concerns with this system of buying and selling human organs an enormous and urgent topic.

3. Catherine Waldby and Robert Mitchell, *Tissue Economies: Blood, Organs, and Cell Lines in Late Capitalism* (Durham: Duke University Press, 2006), 31.

4. Waldby and Mitchell, *Tissue Economies*, 32.

5. Waldby and Mitchell, *Tissue Economies*, 187.

6. Waldby and Mitchell, *Tissue Economies*, 187.

7. See Michael Sappol, *A Traffic of Dead Bodies: Anatomy and Embodied Social Identity in Nineteenth-Century America* (Princeton: Princeton University Press, 2002), 318–319. Sappol does an excellent job of explaining how many of Burke and Hare's American contemporaries rarely suffered the same kinds of legal consequences.

8. Randall Patterson, "The Organ Grinder," *New York* magazine, October 16, 2006, 35.

9. Patterson, "The Organ Grinder," 35.

10. Patterson, "The Organ Grinder," 35.

11. Food and Drug Administration, *Recall of Human Tissue-Biomedical Tissue Services, Ltd.*, October 15, 2005. See https://web.archive.org/web/20170112100149/http://www.fda.gov/BiologicsBloodVaccines/SafetyAvailability/Recalls/ucm053644.htm.

12. Food and Drug Administration, *Human Tissue Recovered by Biomedical Tissue Services, Ltd. (BTS)*, October 26, 2005. See https://web.archive.org/web/20170112170714/http://www.fda.gov/Safety/MedWatch/SafetyInformation/SafetyAlertsforHumanMedicalProducts/ucm152362.htm.

13. Channel 4 in the United Kingdom produced an excellent documentary film on Michael Mastromarino entitled *Body Snatchers of New York*. The program's description from the Channel 4 website gives a strong

sense of how Mastromarino's activities were presented to the public: "The chilling story of New York surgeon Dr. Michael Mastromarino, who illegally carved up hundreds of corpses without permission before selling the bones and tissue on for transplants." The entire documentary can be viewed (where available) online: https://web.archive.org/web/20160522074439/http://www.channel4.com/programmes/bodysnatchers-of-new-york.

14. Office of the District Attorney, Kings County, New York, *Bones for Transplant Taken from Corpses without Consent*, February 23, 2006, 3. See https://web.archive.org/web/20100602011735/http://www.nyc.gov/html/doi/downloads/pdf/tissueharvesting.pdf.

15. Office of the District Attorney, *Bones for Transplant*, 1.

16. Susan Cooke Kittredge, "Black Shrouds and Black Markets," *New York Times*, March 5, 2006, https://www.nytimes.com/2006/03/05/opinion/black-shrouds-and-black-markets.html.

17. Kitty Caparella, "Non-Golfing Judge Set for Body-Parts Case," *Philadelphia Daily News*, August 2, 2008, https://www.philly.com/philly/hp/news_update/20080208_Non-golfing_judge_set_for_body-parts_case.html.

18. Channel 4's documentary *Body Snatchers of New York* offers extensive prison interviews with Michael Mastromarino. See https://web.archive.org/web/20160522074439/http://www.channel4.com/programmes/bodysnatchers-of-new-york.

19. Michele Goodwin, *Black Markets: The Supply and Demand of Body Parts* (Cambridge: Cambridge University Press, 2006), 19. It is worth noting that neither the US House of Representatives nor the Senate has ever shown a strong commitment to changing this loophole. In 2003, Senator Susan Collins of Maine held hearings about strengthening FDA regulation of human tissue products, but no further serious, bipartisan attempts have been made. See Annie Cheney, *Body Brokers: Inside America's Underground Trade in Human Remains* (New York: Broadway Books, 2006), 171.

20. John M. Broder, "In Science's Name, Lucrative Trade in Body Parts," *New York Times*, March 12, 2004, https://www.nytimes.com/2004/03/12/us/in-science-s-name-lucrative-trade-in-body-parts.html.

21. Cheney, *Body Brokers*, xv. Cheney also includes this helpful note: "The prices are valid only for fresh/frozen parts that are used for research and education. The prices may vary depending on the source and the broker. Transportation may or may not be included in the cost."

22. See Michael Sappol, *A Traffic of Dead Bodies: Anatomy and Embodied Social Identity in Nineteenth-Century America* (Princeton: Princeton University Press, 2002), for an extremely thorough history of American body snatching.

23. Kimberly Edds, "UCLA Denies Roles in Cadaver Case," *Washington Post*, March 9, 2004, sec. A, 3.

24. Edds, "UCLA Denies Role in Cadaver Case," sec. A, 3.

25. Goodwin, *Black Markets*, 19.

26. John M. Broder, "U.C.L.A. Halts Donations of Cadavers," *New York Times*, March 10, 2004, https://www.nytimes.com/2004/03/10/us/ucla-halts-donations-of-cadavers-for-research.html.

27. Randal C. Archibold, "2 Accused of Trading in Cadaver Parts," *New York Times*, March 8, 2007, https://www.nytimes.com/2007/03/08/us/08ucla.html.

28. Associated Press, "Plea in Sales of Cadavers by U.C.L.A.," *New York Times*, October 19, 2008, https://www.nytimes.com/2008/10/19/us/19ucla.html.

29. Food and Drug Administration, *Order to Cease Manufacturing and to Retain HCT/Ps—Donor Referral Services*, August 18, 2006. See https://web.archive.org/web/20170505160211/http://www.fda.gov/BiologicsBloodVaccines/SafetyAvailability/TissueSafety/ucm095466.htm.

30. Associated Press, "How a Rogue Body Broker Got Away With It," *NBC.com*, August 28, 2006 (this is a lengthy investigative article on

Guyett's activities and past), http://www.nbcnews.com/id/14518343/ns/health-health_care/t/how-rogue-body-broker-got-away-it.

31. Associated Press, "Body Parts Harvested in N.C. Are Recalled," NBC.com, August 23, 2006, http://www.nbcnews.com/id/14473165/ns/health-health_care/t/body-parts-harvested-nc-are-recalled.

32. Sarah Avery, "Body Tissue Scheme Spurred by Profits," *News and Observer*, October 10, 2009, https://www.mcclatchydc.com/news/crime/article24558994.html.

33. Broder, "In Science's Name, Lucrative Trade in Body Parts," https://www.nytimes.com/2004/03/12/us/in-science-s-name-lucrative-trade-in-body-parts.html.

34. Broder, "In Science's Name, Lucrative Trade in Body Parts," https://www.nytimes.com/2004/03/12/us/in-science-s-name-lucrative-trade-in-body-parts.html.

35. Kate Wilson, Vlad Lavrov, Martina Keller, and Michael Hudson, "Dealer in Human Body Parts Points the Finger," *Sydney Morning Herald*, July 18, 2012, https://www.smh.com.au/politics/federal/dealer-in-human-body-parts-points-the-finger-20120718-229q1.html.

36. See Annie Cheney, "The Resurrection Men: Scenes from the Cadaver Trade," *Harper's Magazine*, March 2004, 45–54. Cheney's article is an excellent piece of reporting on the repurposing of dead bodies. This article predates her book on body brokers.

37. See Mary Roach, *Stiff: The Curious Lives of Human Cadavers* (New York: W. W. Norton, 2003). Roach's very funny and thoughtful book contains numerous examples of dead bodies used in research, for example, for automobile impact studies, for military armor protection studies, and so on. See chap. 1, "A Head is a Terrible Thing to Waste"; chap. 4, "Dead Man Driving"; and chap. 6, "The Cadaver Who Joined the Army."

38. Goodwin, *Black Markets*, 19.

39. See Cheney, *Body Brokers,* 162. Cheney discusses the Lykins case at length.

40. CryoLife is also a company that purchased tissues from another (now closed) tissue company named Lost Mountain Tissues. Lost Mountain Tissues purchased biomaterials from Philip Guyett.

41. State of Minnesota, *Chapter 525A. ANATOMICAL GIFTS,* can be read here: https://www.revisor.mn.gov/statutes/?id=525A.

42. Lawyers representing families in a 2012 court case uncovered documents which show that RTI Biologics in Florida was warned in 2002 not to work with Mastromarino. A lawyer hired by RTI to do a background check strongly advised against taking any biomaterials from Mastromarino. RTI disregarded the advice and continued to extend its contract with him. See Wilson, Lavrov, Keller and Hudson, "Dealer in Human Body Parts Points the Finger," *Sydney Morning Herald,* https://www.smh .com.au/politics/federal/dealer-in-human-body-parts-points-the-finger -20120718-229q1.html.

43. Cheney, "The Resurrection Men," 50.

44. To read the NFDA's full *Best Practices for Organ and Tissue Donation,* see https://web.archive.org/web/20111112215833/http://www.nfda.org/ additional-tools-organtissue/203-organ-and-tissue-donation-best -practices.html.

45. To read the NFDA's policy on tissue and organ donation, see http:// www.nfda.org/component/.../1108-2011-pp-c12-organ-tissue-donation .html.

46. Jonathan Harr, *Funeral Wars* (London: Short Books, 2001), 40.

47. Doug Smith, *Big Death: Funeral Planning in the Age of Corporate Deathcare* (Manitoba: Fernwood, 2007), 56.

48. Smith, *Big Death,* 57.

49. Smith, *Big Death,* 60.

50. Harr, *Funeral Wars,* for the entire legal history.

51. Harr, *Funeral Wars,* 78. This settlement, however, was not the end of the story. What was most legally intriguing about the case was the

Loewen Group's eventual use of a special provision in the North American Free Trade Agreement (NAFTA) called Chapter 11. This is a little known section within NAFTA, and what Chapter 11 stipulates is that any individual or group in one NAFTA nation can sue the government of another NAFTA nation over unfair treatment in the other country's court system. In 1998 the Loewen Group filed a Chapter 11 grievance in order to sue the United States for damages after the O'Keefe trial in Mississippi was complete. The Loewen corporation then proceeded to sue the entire US government for $725 million. Chapter 11 challenges use special tribunals composed of three judges who hear the appeal; the Loewen docket consisted of judges from America, Australia, and Britain. Most significantly, the court systems of either country cannot challenge the three-member tribunal's final decision. The special tribunal's decision is absolute and unstoppable. In June 2003, the Loewen Group lost the Chapter 11 challenge because it had gone through bankruptcy reorganization and reformed in 2002 as Alderwoods, a legally defined American corporation. See Adam Liptak, "NAFTA Tribunals Stir U.S. Worries," *New York Times,* April 18, 2004, for some early reporting on the possibility of NAFTA Chapter 11 cases, https://www.iatp.org/news/nafta-tribunals-stir-us-worries.

52. See the SCI *2007 Purchase Report* for full details about the Alderwoods acquisition: https://web.archive.org/web/20170516224017/http://library.corporate-ir.net/library/10/108/108068/items/283107/SERVICECORPORAT10K.pdf.

53. To read SCI's full *2011 Annual Report,* see http://investors.sci-corp.com/phoenix.zhtml?c=108068&p=irol-reportsAnnual. SCI also expanded, for a time, into other areas of the necroeconomy. In 1996, SCI purchased the UK firm Kenyon Emergency Services, renamed it Kenyon International Emergency Services, and moved its headquarters to Houston. Kenyon does mortuary work and human remains removal after natural and/or man-made disasters such as the 2004 tsunami in Indonesia and India. Thus, in fall 2005, Kenyon received an exclusive, no-bid contract from the Federal Emergency Management Agency (FEMA) to take care of all the dead bodies produced by Hurricane Katrina. I mention Kenyon's no-bid contract in order to highlight the total package of services

offered by necroeconomy conglomerates from, for example, clearing a body off the streets of New Orleans, to identifying the corpse, to finding an SCI-affiliated funeral home for the memorial service, to then locating an SCI-owned cemetery and/or crematorium for final disposition of the remains. Kenyon, however, found FEMA's management of the Gulf Coast cleanup so disorganized that the company broke the contract and started directly working for the State of Louisiana. SCI eventually sold Kenyon in 2007 to focus on other priorities. Kathleen Schalch, "Officials Spar over Katrina Body Recovery," *All Things Considered*, National Public Radio, October 11, 2005, https://www.npr.org/templates/story/story.php ?storyId=4954641.

54. Mary Roach, "Death Wish," *New York Times,* March 11, 2004, https:// www.nytimes.com/2004/03/11/opinion/death-wish.html.

55. Youngner, Anderson, and Schapiro, eds., *Transplanting Human Tissue,* and Stuart J. Youngner, Renee C. Fox, and Laurence J. O'Connell, eds., *Organ Transplantation: Meanings and Realities* (Madison: University of Wisconsin Press, 1996).

56. National Funeral Directors Association, "Organ Donation Agency Hires Funeral Director Liaison," *The Director*, February 2008, 63.

57. See the Carolina Donor Services web page for Funeral Directors: https://www.carolinadonorservices.org/partners/fh-information-and -resources.

58. See the LifeSource Reimbursement page: https://www.life-source.org/ partners/funeral-directors/reimbursement/ and https://www.life-source .org/partners/funeral-directors.

59. Cheney, "The Resurrection Men," 48.

60. In the same *Harper's Magazine* story, a prominent body broker unintentionally described this potentially predatory business scenario to Cheney: "The company would pay ... something like $20,000 for a cadaver, chop it up, and then sell the pieces for $200,000. Poor families would enjoy a new source of income, the company would make a large profit, and the marketplace would finally be provided the parts it desired." Cheney, "The Resurrection Men," 53.

61. Nuffield Council on Bioethics, *Human Bodies: Donations for Medicine and Research*, October 2011, 175. See the full report here: http://www.nuffieldbioethics.org/donation.

62. Henrietta Lacks's story offers a warning about potential biomedical abuses. See Rebecca Skloot, *The Immortal Life of Henrietta Lacks* (New York: Crown, 2010).

63. Waldby and Mitchell, *Tissue Economies*, 187.

Chapter 6

1. The following authors and texts lay the foundations for these concepts: Giorgio Agamben, *Remnants of Auschwitz*, trans. Daniel Heller-Roazen (New York: Zone Books, 1999), 83–84; Michel Foucault, *Society Must Be Defended*, trans. David Macey (New York: Picador, 2003), 248. The concept of necropolitics that I am using here differs from Achille Mbembe's use of necropolitics, a term that he defines as a politics of death. Necropolitics is more accurately, I argue, a political field that involves not the act of death itself but instead the already dead. See Achille Mbembe, *Necropolitics*, trans. Libby Meintjes, *Public Culture* 15, no. 1 (2003): 11–40. Also see *Polygraph 18: Biopolitics, Narrative, Temporality*, issue eds. Rodger Frey and Alexander Ruch (2006).

2. Giorgio Agamben, *Means without End,* trans. Vincenzo Binetti and Cesare Casarino (Minneapolis: University of Minnesota Press, 2000), 3.

3. Agamben, *Means without End,* 4.

4. Agamben, *Means without End*, 6–7.

5. Henry George Liddell and Robert Scott, *A Greek-English Lexicon* (Oxford: Clarendon Press, 1978), 784.

6. Liddell and Scott, *A Greek-English Lexicon,* 1165.

7. Agamben, *Remnants of Auschwitz*, 83.

8. Michel Foucault, *The History of Sexuality, Volume I*, trans. Robert Hurley (New York: Vintage Books, 1978), 136.

9. I also discuss these life and death politics and their impact on the American Personhood Movement in an anthology chapter. See John Troyer, "Defining Personhood to Death," in *A Good Death? Law and Ethics in Practice*, ed. Lynn Hagger and Simon Woods (London: Ashgate Press, 2012), 69–89.

10. Foucault, *Society Must Be Defended*, 248.

11. For an exceptionally good essay on the limits of medical care for individuals who are dying, see Robin Marantz Henig's "Will We Ever Arrive at the Good Death," *New York Times*, August 7, 2005, http://www.nytimes.com/2005/08/07/magazine/07DYINGL.html.

12. Agamben, *Remnants of Auschwitz*, 83–84.

13. Agamben, *Remnants of Auschwitz*, 72.

14. Agamben, *Means without End*, 121.

15. On May 25, 2005, Amnesty International released its annual report on Human Rights. At the press conference for the report's release, Amnesty Secretary General Irene Khan called the US internment camp at Guantanamo Bay "the gulag of our time," which then led to scathing criticism from defenders of the camp and the camp's practices in the United States. The entire episode was a keen example of how the rhetoric of the "camp" in all its forms remains relevant and problematic for sovereign power when confronted by systematic abuses of prisoners. Since that time, reports by journalists from all over the world have highlighted the abuses of power in various US detainment camps. On April 9, 2009, the *New York Review of Books* ran a lengthy article by Mark Danner (*US Torture: Voices from the Black Sites*) that used a report by the International Committee of the Red Cross (ICRC). Both Danner's piece and the ICRC report make clear how US government officials controlled life and ignored death in the camps. See https://www.nybooks.com/articles/2009/04/09/us-torture-voices-from-the-black-sites.

16. Steven H. Miles, "Abu Ghraib: Its Legacy for Military Medicine," *The Lancet* 364 (2004): 725–729.

17. Giorgio Agamben, *Homo Sacer*, trans. Daniel Heller-Roazen (Stanford: Stanford University Press, 1998), 164.

18. Information on the Human Genome Project can be found here: http://www.ornl.gov/sci/techresources/Human_Genome/home.shtml. For early articles about the decoding of the Human Genome in 1999 see I. Dunham et al., "The DNA Sequence of Human Chromosome 22," *Nature* 402 (1999): 489–495, and Peter Little, "The Book of Genes," *Nature* 402 (1999): 467–468.

19. Gilles Deleuze, *Foucault*, trans. Sean Hand (Minneapolis: University of Minnesota Press, 1986), 131.

20. Deleuze, *Foucault*, 132.

21. The problem of visibility with research like the work being done by the Human Genome Project is more complicated than it seems. Oftentimes the research itself will be available for public view, but it's either not reported on in a significant way by the popular press and/or most of the general public does not know how to digest the information. The largest stumbling block when explaining these changes to the human for a general audience is the sheer science-fiction-sounding quality of the work.

22. Deleuze, *Foucault*, 132.

23. This specific President's Council on Bioethics was created by President George W. Bush on November 28, 2001, by Executive Order 13237. President Jimmy Carter formed the original President's Council in the late 1970s. The mission of the President's Council on Bioethics, per President Bush's executive order, was as follows:

> The Council shall advise the President on bioethical issues that may emerge as a consequence of advances in biomedical science and technology. In connection with its advisory role, the mission of the Council includes the following functions: 1. to undertake fundamental inquiry into the human and moral significance of developments in biomedical and behavioral science and technology; 2. to explore specific ethical and policy questions related to these developments; 3. to provide a forum for a national discussion of bioethical issues; 4. to facilitate a greater understanding of bioethical issues; and 5. to explore possibilities for useful international collaboration on bioethical issues.

On November 24, 2009, President Barack Obama created a new group, the Presidential Commission for the Study of Bioethical Issues, which replaced the President's Council on Bioethics. Information on both groups can be found at www.bioethics.gov. President Donald Trump has not yet convened his own Presidential Council on Bioethics.

24. United States, The President's Council on Bioethics, *Beyond Therapy: Biotechnology and the Pursuit of Happiness* (Washington, DC: Government Printing Office, 2003), 161.

25. This is taken from "The Transhumanist Declaration" at the World Transhumanist Association website, http://www.transhumanism.org. The quotation is the first of the eight-part declaration. Here are all eight parts of the declaration:

1.) Humanity stands to be profoundly affected by science and technology in the future. We envision the possibility of broadening human potential by overcoming aging, cognitive shortcomings, involuntary suffering, and our confinement to planet Earth. 2.) We believe that humanity's potential is still mostly unrealized. There are possible scenarios that lead to wonderful and exceedingly worthwhile enhanced human conditions. 3.) We recognize that humanity faces serious risks, especially from the misuse of new technologies. There are possible realistic scenarios that lead to the loss of most, or even all, of what we hold valuable. Some of these scenarios are drastic, others are subtle. Although all progress is change, not all change is progress. 4.) Research effort needs to be invested into understanding these prospects. We need to carefully deliberate how best to reduce risks and expedite beneficial applications. We also need forums where people can constructively discuss what should be done, and a social order where responsible decisions can be implemented. 5.) Reduction of existential risks, and development of means for the preservation of life and health, the alleviation of grave suffering, and the improvement of human foresight and wisdom should be pursued as urgent priorities, and heavily funded. 6.) Policy making ought to be guided by responsible and inclusive moral vision, taking seriously both opportunities and risks, respecting autonomy and individual rights, and showing solidarity with and concern for the interests and dignity of all people around the globe. We must also consider our moral responsibilities towards generations that will exist in the future. 7.) We advocate the well-being of all sentience, including humans, non-human animals, and any future artificial intellects, modified life forms, or other intelligences to which technological and scientific advance may give rise. 8.) We favour allowing individuals wide personal choice over how they enable their lives. This includes use of techniques that may be developed to assist memory, concentration, and mental energy; life extension therapies; reproductive choice technologies; cryonics procedures; and many other possible human modification and enhancement technologies.

26. See Mark O'Connell, *To Be a Machine: Adventures among Cyborgs, Utopians, Hackers, and the Futurists Solving the Modest Problem of Death* (London: Granta Books, 2017). O'Connell's book won the 2018 Wellcome Book Prize.

27. Steve Lohr, "Just How Old Can He Go?" *New York Times*, December 27, 2004, https://www.nytimes.com/2004/12/27/technology/just-how -old-can-he-go.html. Ray Kurzweil's own books, *The Age of Spiritual Machines* (1999), *Fantastic Voyage: Live Long Enough to Live Forever* (2004), and *The Singularity Is Near: When Humans Transcend Biology* (2005), articulate how cutting-edge technology can significantly extend human life.

28. See Elaine Kasket, *All the Ghosts in the Machine: Illusions of Immortality in the Digital Age* (London: Robinson, 2019).

29. The President's Council on Bioethics, *Beyond Therapy,* 190–191.

30. The case of Terri Schiavo in 2005 offered a momentary glimpse into how the politics of dying can play out in a hyperpoliticized context. The entire Terri Schiavo case is a long and complicated story that went in political, bioethical, and medical directions no one person seemed to ever anticipate. A vast number of points have been raised elsewhere about the role of sovereign power, theology, contemporary American party politics, and the definition of death in American law. For a complete and extremely thorough time line of the Schiavo case, go to the following website run by the University of Miami Ethics Programs and the Shepard Broad Law Center at Nova Southeastern University: http:// www.miami.edu/ethics/schiavo/timeline.htm.

31. Georges Canguilhem, *The Normal and the Pathological*, trans. Carolyn R. Fawcett (New York: Zone Books, 1989), 236.

32. Agamben, *The Open*, 80.

Chapter 7

1. Taken from the Patent and Trademark Office website: http://www .uspto.gov.

2. Rick Weiss, "U.S. Denies Patent for a Too-Human Hybrid," *Washington Post,* February 13, 2005, sec. A, 3.

3. Weiss, "U.S. Denies Patent for a Too-Human Hybrid," sec. A, 3.

4. Weiss, "U.S. Denies Patent for a Too-Human Hybrid," sec. A, 3.

5. The biopolitical concerns raised by Newman's patent application also helpfully illustrate broader human insecurity about the sanctity of *Homo sapiens* both as a concept and as an animal species. Concerns about altering the human body and producing "hybrids" seems to erroneously suggest that those kinds of things are not already happening within the gene pool or, at least, in *Body Worlds* exhibitions. What really begins to emerge from these overt concerns is precisely the fear of a new kind of "human monster." Assuring individual and civil rights to any and all such monsters seems more important for public discussion than whether or not humans are being experimented into oblivion. That new kind of legal status would of course mean changing the juridical definition of the "person," and more likely than not it would mean altering the "human" as a scientific category.

6. The entire Supreme Court decision in *Diamond v. Chakrabarty*, 447 U.S. 303 (1980), can be read here: https://caselaw.findlaw.com/us-supreme -court/447/303.html.

7. Jeremy Rifkin, *The Biotech Century* (New York: Jeremy P. Tarcher/ Putnam, 1998), 42.

8. *Diamond v. Chakrabarty,* https://caselaw.findlaw.com/us-supreme -court/447/303.html.

9. *Diamond v. Chakrabarty.* In the majority decision written by Chief Justice Warren E. Burger, one section of the decision acknowledges these concerns when discussing friend-of-the-court briefs filed on behalf of Diamond:

Scientists, among them Nobel laureates, are quoted suggesting that genetic research may pose a serious threat to the human race, or, at the very least, that the dangers are far too substantial to permit such research to proceed apace at this time. We are told that genetic research and related technological developments may spread pollution and disease, that it may result in a loss of genetic diversity, and that its practice may tend to depreciate the value of human life. These arguments are forcefully, even passionately, presented; they remind us that, at times, human ingenuity seems unable to control fully the forces it creates—that, with Hamlet, it is sometimes better "to bear those ills we have than fly to others we know not of." (https://caselaw.findlaw.com/us-supreme -court/447/303.html)

10. Title 35 United States Code 101 regarding patentable inventions states, "Whoever invents or discovers any new and useful process, machine, manufacture, or composition of matter, or any new and useful improvement thereof, may obtain a patent therefor, subject to the conditions and requirements of this title." See https://www.law.cornell.edu/ uscode/text/35/101.

11. Robert H. Blank, "Technology and Death Policy: Redefining Death," *Mortality* 6, no. 2 (2002): 191–202.

12. Giorgio Agamben, *Homo Sacer*, trans. Daniel Heller-Roazen (Stanford: Stanford University Press, 1998), 163.

13. James J. Hughes, "The Death of Death," *Advances in Experimental Medicine and Biology* 550 (2004): 80.

14. Donna Haraway's work on the human body's relationship with technology and science is extremely relevant here. See *Simians, Cyborgs, and Women: The Reinvention of Nature* (New York: Routledge, 1991) and *Modest_Witness@Second_Millennium.FemaleMan_Meets_OncoMouse: Feminism and Technoscience* (New York: Routledge, 1997).

15. Antonio Regalado, "A Startup Is Pitching a Mind-Uploading Service that Is '100 Percent Fatal'" *MIT Technology Review*, March 13, 2018, https://www.technologyreview.com/s/610456/a-startup-is-pitching-a -mind-uploading-service-that-is-100-percent-fatal.

16. Hughes, "The Death of Death," 80.

17. Raymond Williams, *The Politics of Modernism: Against the New Conformists* (New York: Verso, 1989), 134.

18. Michel Foucault, *The History of Sexuality, Volume I*, trans. Robert Hurley (New York: Vintage Books, 1990), 159.

19. Michel Foucault, *The Order of Things*, trans. Alan Sheridan (New York: Vintage Books, 1970), 387.

Works Cited and Consulted

Agamben, Giorgio. *Homo Sacer*. Translated by Daniel Heller-Roazen. Stanford: Stanford University Press, 1998.

Agamben, Giorgio. *Means without End*. Translated by Vincenzo Binetti and Cesare Casarino. Minneapolis: University of Minnesota Press, 2000.

Agamben, Giorgio. *The Open*. Translated by Kevin Atell. Stanford: Stanford University Press, 2004.

Agamben, Giorgio. *Remnants of Auschwitz*. Translated by Daniel Heller-Roazen. New York: Zone Books, 1999.

Agence France-Presse. "Copulating Corpses Raise the Roof in Berlin." *AFP*, May 7, 2009. https://www.france24.com/en/20090507-exhibition -germany-doctor-death-copulating-corpses-raise-roof-berlin-hagens -anatomy.

Anderson, Martha W., and Renie Schapiro. "From Donor to Recipient: The Pathway and Business of Donated Tissues." In *Transplanting Human Tissue: Ethics, Policy, and Practice*, edited by Stuart J. Youngner, Martha W. Anderson, and Renie Schapiro, chapter 1. Oxford: Oxford University Press, 2004.

Archibold, Randal C. "2 Accused of Trading in Cadaver Parts." *New York Times*, March 8, 2007. https://www.nytimes.com/2007/03/08/us/08ucla .html.

Associated Press. "Body Parts Harvested in N.C. Are Recalled." *NBC .com*, August 23, 2006. http://www.nbcnews.com/id/14473165/ns/health -health_care/t/body-parts-harvested-nc-are-recalled.

Associated Press. "How a Rogue Body Broker Got Away With It." *NBC. com*, August 28, 2006. http://www.nbcnews.com/id/14518343/ns/health -health_care/t/how-rogue-body-broker-got-away-it.

Associated Press. "Plea in Sales of Cadavers by U.C.L.A." *New York Times*, October 19, 2008. https://www.nytimes.com/2008/10/19/us/19ucla .html.

Avery, Sarah. "Body Tissue Scheme Spurred by Profits." *News and Observer*, October 10, 2009. https://www.mcclatchydc.com/news/crime/ article24558994.html.

Barnes, Carl Lewis. *The Art and Science of Embalming*. Chicago: Trade Periodical, 1896.

Barnes, Carl Lewis. *Barnes School of Anatomy, Sanitary Science and Embalming*. New York: [self-published], 1905.

Barnes, Carl Lewis. Public Lecture. Annual Convention of the Connecticut Funeral Directors Association. September 12–13, 1905.

Barthes, Roland. *Camera Lucida*. New York: Hill and Wang, 1981.

Becker, Ernest. *The Denial of Death*. New York: Free Press, 1973.

Bedino, James H. *AIDS: A Comprehensive Update for Embalmers*. Research and Education Department of The Champion Company, No. 616 (1993).

Benjamin, Walter. *Illuminations*. Translated by Harry Zohn. New York: Schocken Books, 1968.

Benjamin, Walter. "A Short History of Photography." In *Classic Essays on Photography*, edited by Alan Trachtenberg, 199–217. New Haven: Leete Island Books, 1980.

Bersani, Leo. *Homos*. Cambridge: Harvard University Press, 1995.

Bisga Fluid Advertisement. *Casket*, December 1902, 80–81.

Bisga Fluid advertisement. *Sunnsyside*, October 1902, 5.

Blank, Robert H. "Technology and Death Policy: Redefining Death." *Mortality* 6, no. 2 (2002): 191–202.

Broder, John M. "In Science's Name, Lucrative Trade in Body Parts." *New York Times*, March 12, 2004. https://www.nytimes.com/2004/03/12/us/in-science-s-name-lucrative-trade-in-body-parts.html.

Broder, John M. "U.C.L.A. Halts Donations of Cadavers." *New York Times*, March 10, 2004. https://www.nytimes.com/2004/03/10/us/ucla-halts-donations-of-cadavers-for-research.html.

Burns, Stanley. *Sleeping Beauties: Memorial Photography in America*. New York: Burns Archive Press, 1990.

Burns, Stanley. *Sleeping Beauty II: Grief, Bereavement and the Family in Memorial Photography*. New York: Burns Archive Press, 2002.

Butler, Judith. *Bodies that Matter*. New York: Routledge, 1993.

Byrne, Paul A., and Walter F. Weaver. "'Brain Death' Is Not Death." *Advances in Experimental Medicine and Biology* 550 (2004): 43–49.

Canguilhem, Georges. *The Normal and the Pathological*. Translated by Carolyn R. Fawcett. New York: Zone Books, 1989.

Cantor, Norman. *After We Die: The Life and Times of The Human Cadaver*. Washington, DC: Georgetown University Press, 2010.

Caparella, Kitty. "Non-Golfing Judge Set for Body-Parts Case." *Philadelphia Daily News*, August 2, 2008. https://www.philly.com/philly/hp/news_update/20080208_Non-golfing_judge_set_for_body-parts_case.html.

Carney, Scott. "Why a Kidney (Street Value $3,000) Sells for $85,000." *Wired*, May 8, 2007. https://www.wired.com/2007/05/india-transplants-prices.

Charo, Alta. "Dusk, Dawn, and Defining Death: Legal Classifications and Biological Categories." In *The Definition of Death: Contemporary*

Controversies, edited by Stuart J. Youngner, Robert M. Arnold, and Renie Schapiro, 277–292. Baltimore: Johns Hopkins University Press, 1999.

Cheney, Annie. *Body Brokers: Inside America's Underground Trade in Human Remains.* New York: Broadway Books, 2006.

Cheney, Annie. "The Resurrection Men: Scenes from the Cadaver Trade." *Harper's Magazine,* March 2004, 45–54.

Connolly, Kate. "Fury at Exhibit of Corpses Having Sex." *Guardian,* May 6, 2009. https://www.theguardian.com/world/2009/may/06/german-artist -sex-death.

Cooke Kittredge, Susan. "Black Shrouds and Black Markets." *New York Times,* March 5, 2006. https://www.nytimes.com/2006/03/05/opinion/ black-shrouds-and-black-markets.html.

Crary, John. *Techniques of the Observer.* Cambridge: MIT Press, 1991.

Danner, Mark. "US Torture: Voices from the Black Sites." *New York Review of Books,* April 9, 2009. https://www.nybooks.com/articles/2009/04/09/ us-torture-voices-from-the-black-sites.

De Certeau, Michel. *Heterologies: Discourse on the Other.* Translated by Brian Massumi. Minneapolis: University of Minnesota Press, 1986.

De Certeau, Michel. *The Practice of Everyday Life.* Translated by Steven Rendall. Berkeley: University of California Press, 1984.

De Certeau, Michel. *The Writing of History.* Translated by Tom Conley. New York: Columbia University Press, 1988.

Deleuze, Gilles. *Foucault.* Translated by Sean Hand. Minneapolis: University of Minnesota Press, 1986.

Deleuze, Gilles, and Félix Guattari. *A Thousand Plateaus.* Translated by Brian Massumi. Minneapolis: University of Minnesota Press, 1987.

Diamond v. Chakrabarty, 447 U.S. 303 (1980). https://caselaw.findlaw .com/us-supreme-court/447/303.html.

Donaldson, Peter J. "Denying Death: A Note Regarding Ambiguities in the Current Discussion." *Omega,* November 1972, 285–290.

Dumont, Richard G., and Dennis C. Foss. *The American View of Death: Acceptance or Denial?* Cambridge: Schenkman, 1972.

Duncan, David Ewing. "How Long Do You Want to live?" *New York Times,* August 25, 2012. https://nytimes.com/2012/08/26/sunday-review/how -long-do-you-want-to-live.html.

Dunham, I., et al. "The DNA Sequence of Human Chromosome 22." *Nature* 402 (1999): 489–495.

Edds, Kimberly. "UCLA Denies Roles in Cadaver Case." *Washington Post,* March 9, 2004, sec. A, 3.

Farrell, James. *Inventing the American Way of Death, 1830–1912.* Philadelphia: Temple University Press, 1980.

Foucault, Michel. *Abnormal: Lectures at the Collège de France, 1974–1975.* Translated by Graham Burchell. New York: Picador, 2003.

Foucault, Michel. *Birth of the Clinic.* Translated by Alan Sheridan. New York: Vintage, 1973.

Foucault, Michel. *Ethics, Subjectivity, and Truth.* Edited by Paul Rabinow. Translated by Robert Hurley. New York: New Press, 1997.

Foucault, Michel. *The History of Sexuality, Volume I.* Translated by Robert Hurley. New York: Vintage Books, 1978.

Foucault, Michel. *The History of Sexuality, Volume III.* Translated by Robert Hurley. New York: Vintage Books, 1986.

Foucault, Michel. *The Order of Things.* Translated by Alan Sheridan. New York: Vintage Books, 1970.

Foucault, Michel. *Society Must Be Defended.* Translated by David Macey. New York: Picador, 2003.

Frederick, Jerome F. "AIDS—Identification and Preparation." *The Director,* July 1985, 8–11, 43–44.

Frey, Rodger, and Alexander Ruch, eds. *Polygraph 18: Biopolitics, Narrative, Temporality.* Durham: Duke University Press, 2006.

Funeral Directors Services Association of Greater Chicago Infectious/Contagious Disease Committee. "AIDS Update." *The Director,* January 1993, 56–57.

Gilligan, T. Scott, and Thomas F. H. Stueve. *Mortuary Law.* 9th ed. Cincinnati: Cincinnati Foundation for Mortuary Education, 1995.

Goodwin, Michele. *Black Markets: The Supply and Demand of Body Parts.* Cambridge: Cambridge University Press, 2006.

Gunning, Tom. "Phantom Images and Modern Manifestations." In *Fugitive Images: From Photography to Video,* edited by Patrice Petro, 42–71. Bloomington: Indiana University Press, 1995.

Gunning, Tom. "Tracing the Individual Body: Photography, Detectives, and Early Cinema." In *Cinema and the Invention of Modern Life,* edited by Leo Charney and Vanessa R. Schwartz, 15–45. Berkeley: University of California Press, 1995.

Habenstein, Robert W., and William M. Lamers. *The History of American Funeral Directing.* 4th ed. Milwaukee: National Funeral Directors Association of the United States, 1996.

Haraway, Donna. *Modest_Witness@Second_Millennium.FemaleMan_Meets_OncoMouse: Feminism and Technoscience.* New York: Routledge, 1997.

Haraway, Donna. *Simians, Cyborgs, and Women: The Reinvention of Nature.* New York: Routledge, 1991.

Harr, Jonathan. *Funeral Wars.* London: Short Books, 2001.

Hearn, Michael. "Photographs and Memories." *The Director,* January 1992, 10–13, 57–58.

Henig, Robin Marantz. "Will We Ever Arrive at the Good Death." *New York Times,* August 7, 2005. http://www.nytimes.com/2005/08/07/magazine/07DYINGL.html.

Hughes, James J. "The Death of Death." *Advances in Experimental Medicine and Biology,* no. 550 (2004): 79–87.

Institute for Plastination. *Donating Your Body for Plastination.* 7th rev. ed. Heidelberg, Germany: Institute for Plastination, 2004.

Itzkoff, Dave. "Cadaver Sex Exhibition in Germany Is Criticized." *New York Times,* May 7, 2009. https://www.nytimes.com/2009/05/08/arts/design/08arts-CADAVERSEXEX_BRF.html.

Jameson, Fredric. *Postmodernism.* Durham: Duke University Press, 1991.

Kasket, Elaine. *All the Ghosts in the Machine: Illusions of Immortality in the Digital Age.* London: Robinson, 2019.

Kübler-Ross, Elisabeth. *On Death and Dying.* New York: Macmillan, 1969.

Kurzweil, Ray. *The Age of Spiritual Machines.* New York: Penguin, 1999.

Kurzweil, Ray. *Fantastic Voyage: Live Long Enough to Live Forever.* New York: Plume Books, 2004.

Kurzweil, Ray. *The Singularity Is Near: When Humans Transcend Biology.* New York: Viking Press, 2005.

Kykkotis, I. *English-Modern Greek and Modern Greek-English Dictionary.* London: Percy Lund, Humphries, 1957.

Laderman, Gary. *Rest in Peace: A Cultural History of Death and the Funeral Home in Twentieth-Century America.* New York: Oxford University Press, 2003.

Lensing, Michael. "Arrangement Conference for AIDS Related Deaths." *The Director,* December 1996, 6–8.

Leppert, Richard. *Art and the Committed Eye: The Cultural Functions of Imagery.* Boulder, CO: Westview Press, 1996.

Lesy, Michael. *Wisconsin Death Trip.* New York: Pantheon Books, 1973.

Liddell, Henry George, and Robert Scott. *A Greek-English Lexicon.* Oxford: Clarendon Press, 1978.

Liptak, Adam. "NAFTA Tribunals Stir U.S. Worries." *New York Times,* April 18, 2004. https://www.iatp.org/news/nafta-tribunals-stir-us-worries.

Little, Peter. "The Book of Genes." *Nature* 402 (1999): 467–468.

Lizza, John P. "Defining Death for Persons and Human Organisms." *Theoretical Medicine and Bioethics* 20, no. 5 (1999): 439–453.

Lofland, Lyn. *The Craft of Dying: The Modern Face of Death.* 40th anniversary ed. Cambridge: MIT Press, 2019.

Lohr, Steve. "Just How Old Can He Go?" *New York Times*, December 27, 2004. https://www.nytimes.com/2004/12/27/technology/just-how-old-can-he-go.html.

Lyotard, Jean-François. *The Postmodern Condition: A Report on Knowledge.* Translated by Geoff Bennington and Brian Massumi. Minneapolis: University of Minnesota Press, 1984.

Mayer, Robert G. *Embalming: History, Theory and Practice.* 3rd ed. New York: McGraw-Hill, 2000.

Mayer, Robert G. "Offering a Traditional Funeral to All Families." *The Director*, September 1987, 28–30.

Mbembe, Achille. "Necropolitics." Translated by Libby Meintjes. *Public Culture* 15, no. 1 (2003): 11–40.

Miles, Steven H. "Abu Ghraib: Its Legacy for Military Medicine." *The Lancet*, no. 364 (2004): 725–729.

Minnesota State Government. *Chapter 525A. ANATOMICAL GIFTS.* St. Paul: Minnesota State Legislature. https://www.revisor.mn.gov/statutes/?id=525A.

National Commission on Acquired Immune Deficiency Syndrome. "America Living with AIDS." *The Director*, January 1992, 18–22.

National Funeral Directors Association. "AIDS Precautions for Funeral Service Personnel and Others." *The Director*, June 1985, 1.

National Funeral Directors Association. *Best Practices for Organ and Tissue Donation.* Milwaukee: National Funeral Directors Association, 2011. https://web.archive.org/web/20111112215833/http://www.nfda.org/additional-tools-organtissue/203-organ-and-tissue-donation-best-practices.html.

National Funeral Directors Association. "HIV on the Job." *The Director,* January 1992, 16–17.

National Funeral Directors Association. "Organ Donation Agency Hires Funeral Director Liaison." *The Director,* February 2008, 63.

National Funeral Directors Association. *Policy on Tissue and Organ Donation.* Milwaukee: National Funeral Directors Association, 2011. http://www.nfda.org/component/.../1108-2011-pp-c12-organ-tissue-donation.html.

National Funeral Directors Association Board of Governors. "Acquired Immune Deficiency Syndrome Policy of the National Funeral Directors Association of the United States, Inc." *The Director,* October 1985, 19.

New York State Government. Office of the District Attorney. *Bones for Transplant Taken from Corpses without Consent.* New York: Kings County, February 23, 2006. https://web.archive.org/web/20100602011735/http://www.nyc.gov/html/doi/downloads/pdf/tissueharvesting.pdf.

Nuffield Council on Bioethics. *Human Bodies: Donations for Medicine and Research,* October 2011, 175. http://www.nuffieldbioethics.org/donation.

Nunokawa, Jeff. "'All the Sad Young Men': AIDS and the Work of Mourning." *Yale Journal of Criticism* 4, no. 2 (1991): 1–12.

O'Connell, Mark. *To Be a Machine: Adventures among Cyborgs, Utopians, Hackers, and the Futurists Solving the Modest Problem of Death.* London: Granta Books, 2017.

Patterson, Randall. "The Organ Grinder." *New York* magazine, October 16, 2006, 35.

"Proceedings of the 3rd Annual Meeting of Association of State and Provincial Boards of Health and Embalmers' Examining Boards of North America." Chicago, IL, 1906.

"Proceedings of the 4th Annual Joint Conference of Embalmers' Examining Boards of North America." Norfolk, VA, 1907.

"Proceedings of the 5th Annual Joint Conference of Embalmers' Examining Boards of North America." Washington, DC, 1908.

"Proceedings of the 6th Annual Joint Conference of Embalmers' Examining Boards of North America." Louisville, KY, 1909.

"Proceedings of the 7th Annual Joint Conference of Embalmers' Examining Boards of North America." Chicago, IL, 1910.

Rabinow, Paul. *Essays on the Anthropology of Reason*. Princeton: Princeton University Press, 1996.

Regalado, Antonio. "A Startup Is Pitching a Mind-Uploading Service that Is '100 Percent Fatal.'" *MIT Technology Review*, March 13, 2018. https://www.technologyreview.com/s/610456/a-startup-is-pitching-a -mind-uploading-service-that-is-100-percent-fatal.

Reuters. "Copulating Corpses Spark Outrage in Berlin Show." *Reuters*, May 6, 2009. https://www.reuters.com/article/us-finearts-corpses/copulating -corpses-spark-outrage-in-berlin-show-idUSTRE5455CI20090506.

Rifkin, Jeremy. *The Biotech Century*. New York: Jeremy P. Tarcher/ Putnam, 1998.

Roach, Mary. "Death Wish." *New York Times,* March 11, 2004. https:// www.nytimes.com/2004/03/11/opinion/death-wish.html.

Roach, Mary. *Stiff: The Curious Lives of Human Cadavers*. New York: W. W. Norton, 2003.

Ruby, Jay. *Secure the Shadow*. Cambridge: MIT Press, 1995.

Sanders, Dalton. "Err on the Side of Caution." *The Director*, April 1997, 73–74.

Sappol, Michael. *A Traffic of Dead Bodies: Anatomy and Embodied Social Identity in Nineteenth-Century America*. Princeton: Princeton University Press, 2004.

Schalch, Kathleen. "Officials Spar over Katrina Body Recovery," *All Things Considered,* National Public Radio. Washington, DC: NPR, October 11, 2005. https://www.npr.org/templates/story/story.php?storyId=4954641.

Schivelbusch, Wolfgang. *The Railway Journey: The Industrialization of Time and Space in the 19th Century.* Berkeley: University of California Press, 1977.

Schwartz, Vanessa. "Cinematic Spectatorship before the Apparatus: The Public Taste for Reality in *Fin-de-Siècle* Paris." In *Cinema and the Invention of Modern Life,* edited by Leo Charney and Vanessa R. Schwartz, 297–319. Berkeley: University of California Press, 1995.

Service Corporation International. *2007 Alderwoods Purchase Report.* https://web.archive.org/web/20170516224017/http://library.corporate -ir.net/library/10/108/108068/items/283107/SERVICECORPORAT10K .pdf.

Service Corporation International. *2011 Annual Report.* http://investors .sci-corp.com/phoenix.zhtml?c=108068&p=irol-reportsAnnual.

Shewmon, D. Alan, and Elizabeth Seitz Shewmon. "The Semiotics of Death and Its Medical Implications." *Advances in Experimental Medicine and Biology,* no. 550 (2004): 89–114.

Simmel, Georg. *The Sociology of Georg Simmel.* Translated by Kurt H. Wolff. New York: Free Press, 1950.

Singer, Ben. "Modernity, Hyperstimulation, and the Rise of Popular Sensationalism." In *Cinema and the Invention of Modern Life,* edited by Leo Charney and Vanessa R. Schwartz, 72–99. Berkeley: University of California Press, 1995.

Skloot, Rebecca. *The Immortal Life of Henrietta Lacks.* New York: Crown, 2010.

Smith, Doug. *Big Death: Funeral Planning in the Age of Corporate Death-care.* Manitoba: Fernwood, 2007.

Spinoza, Benedict de. *Ethics.* Translated by G. H. R. Parkinson. London: Oxford University Press, 2000.

Strub, Clarence G., and L. G. "Darko" Frederick. *The Principles and Practice of Embalming.* 5th ed. Dallas: Professional Training Schools, 1989.

Troyer, John. "Defining Personhood to Death." In *A Good Death? Law and Ethics in Practice*, edited by Lynn Hagger and Simon Woods, 69–89. London: Ashgate Press, 2012.

United States. Food and Drug Administration. *Human Tissue Recovered by Biomedical Tissue Services, Ltd. (BTS)*. Washington, DC: Department of Health and Human Services, October 26, 2005. https://web.archive.org/web/20170112170714/http://www.fda.gov/Safety/MedWatch/SafetyInformation/SafetyAlertsforHumanMedicalProducts/ucm152362.htm.

United States. Food and Drug Administration. *Order to Cease Manufacturing and to Retain HCT/Ps—Donor Referral Services*. Washington, DC: Department of Health and Human Services, August 18, 2006. https://web.archive.org/web/20170505160211/http://www.fda.gov/BiologicsBloodVaccines/SafetyAvailability/TissueSafety/ucm095466.htm.

United States. Food and Drug Administration. *Recall of Human Tissue-Biomedical Tissue Services, Ltd*. Washington, DC: Department of Health and Human Services, October 13, 2005. https://web.archive.org/web/20170112100149/http://www.fda.gov/BiologicsBloodVaccines/SafetyAvailability/Recalls/ucm053644.htm.

United States. President's Commission for the Study of Ethical Problems in Medicine and Biomedical Behavioral Research. *Defining Death: A Report on the Medical, Legal and Ethical Issues in the Determination of Death*. Washington, DC: Government Printing Office, 1981.

United States. The President's Council on Bioethics. *Beyond Therapy: Biotechnology and the Pursuit of Happiness*. Washington, DC: Government Printing Office, 2003.

United States. The President's Council on Bioethics. *Controversies in the Determination of Death: A White Paper by the President's Council on Bioethics*. Washington, DC: Government Printing Office, 2009.

University of Minnesota Center for Bioethics. "Determination of Death: Reading Packet on the Determination of Death." Minneapolis: University of Minnesota, 1997.

Van Der Zee, James. *The Harlem Book of the Dead*. Dobbs Ferry, NY: Morgan & Morgan, 1978.

Virno, Paolo. *A Grammar of the Multitude*. Translated by Isabella Bertoletti, James Cascaito, and Andrea Casson. New York: Semiotext[e], 2004.

Von Hagens, Gunther. "Body Worlds Sex Couple: The Debate." *London Evening Standard*, June 23, 2009. http://www.thisislondon.co.uk/standard -home/body-worlds-sex-couple-the-debate-6801712.html.

Von Hagens, Gunther. "Cadaver Exhibits Are Part Science, Part Sideshow," *All Things Considered*, National Public Radio. Washington, DC: NPR, August 10, 2006. http://www.npr.org/templates/story/story.php?storyId =5553329.

Von Hagens, Gunther. *KÖRPERWELTEN Exhibition Guide*. 4th ed. Heidelberg: Institute for Plastination, 2001.

Waldby, Catherine. *AIDS and the Body Politic*. New York: Routledge, 1996.

Waldby, Catherine, and Robert Mitchell. *Tissue Economies: Blood, Organs, and Cell Lines in Late Capitalism*. Durham: Duke University Press, 2006.

Walter, Tony. "Body Worlds: Clinical Detachment and Anatomical Awe." *Sociology of Health & Illness* 26, no. 4 (2004): 464–488.

Watney, Simon. *Imagine Hope: AIDS and Gay Identity*. London: Routledge, 2000.

Watney, Simon. *Policing Desire: Pornography, AIDS and the Media*. Minneapolis: University of Minnesota, 1996.

Weiss, Rick. "U.S. Denies Patent for a Too-Human Hybrid." *Washington Post*, February 13, 2005, sec. A, 3.

Weston, Kath. *Families We Choose: Lesbians, Gays, Kinship*. New York: Columbia University Press, 1991.

Williams, Raymond. *The Politics of Modernism: Against the New Conformists*. New York: Verso, 1989.

Wilson, Kate, Vlad Lavrov, Martina Keller, and Michael Hudson. "Dealer in Human Body Parts Points the Finger." *Sydney Morning Herald*, July 18, 2012. https://www.smh.com.au/politics/federal/dealer-in-human-body -parts-points-the-finger-20120718-229q1.html.

Youngner, Stuart J., Martha W. Anderson, and Renie Schapiro, eds. *Transplanting Human Tissue: Ethics, Policy, and Practice.* Oxford: Oxford University Press, 2004.

Youngner, Stuart J., Robert M. Arnold, and Renie Schapiro, eds. *The Definition of Death: Contemporary Controversies.* Baltimore: Johns Hopkins University Press, 1999.

Youngner, Stuart J., Renee C. Fox, and Laurence J. O'Connell, eds. *Organ Transplantation: Meanings and Realities.* Madison: University of Wisconsin Press, 1996.

Index